Left to Our Own Devices

T0323312

Left to Our Own Devices

Coping with Insecure Work in a Digital Age

JULIA TICONA

OXFORD
UNIVERSITY PRESS

OXFORD
UNIVERSITY PRESS

Oxford University Press is a department of the University of Oxford. It furthers
the University's objective of excellence in research, scholarship, and education
by publishing worldwide. Oxford is a registered trade mark of Oxford University
Press in the UK and certain other countries.

Published in the United States of America by Oxford University Press
198 Madison Avenue, New York, NY 10016, United States of America.

Library of Congress Control Number: 2021044706
ISBN 978–0–19–763100–3 (pbk.)
ISBN 978–0–19–069128–8 (hbk.)

DOI: 10.1093/oso/9780190691288.001.0001

1 3 5 7 9 8 6 4 2

Paperback printed by LSC Communications, United States of America
Hardback printed by Bridgeport National Bindery, Inc., United States of America

Contents

Preface

During high school, my friends and I would leave school in our small city and head to low-level service jobs at fast-food restaurants and retail stores. Our phones—which at that point were much less "smart" than they are now—were our lifelines to break up the boredom, complain about customers, and calculate our paychecks. Once we could drive, they gave us a sense of security as we drove along ice-crusted roads and helped us spread the word about after-work parties in the fields along poorly marked country roads.

After our shifts, phones were shoved in glove compartments, abandoned in back seats, and stashed in bags. A few friends, anticipating calls from harried managers or coworkers asking them to come in for early shifts, or their parents' annoyed inquiries about when they planned to come home, turned their phones off. Network coverage was spotty out around the city limits, and a bad connection would send callers straight to voicemail with minimal ringing; conveniently, so did turning the phone off. "I swear I didn't even see it! No service!" was an excuse that saved my friendships with coworkers on more than one occasion.

As I attended college, at an elite school far from my hometown, a new set of friends secured coveted internships at investment banks and politicians' offices in big cities, and I noticed them developing a much different relationship to these same devices. Pings and buzzes pulled my friends away from social events, woke them from sleep, and disciplined them to a new way of working. Their jobs were stepping stones to the next opportunity, which confusingly only required them to work even harder and longer than the step before. As we graduated amid the wreckage of the Great Recession, we were told job markets were tight and that we should defensively curate every wrinkle of our online presence. Suddenly, the sites that hosted our memories were also the very things that could stand between us and our already precarious futures; opportunities were thin, so we obediently complied.

Back home, manufacturing plants and hospitals closed, and full-time work evaporated. Juggling part-time school and certification programs, my friends struggled to scrape together part-time work. While picking up a prescription at a chain pharmacy in town, I noticed the bulky hiring kiosk, where job seekers filled out online applications, was plastered with an "Out of Order" sign that directed prospective workers to visit a website to fill out the application instead. Having never subscribed to the slow and unreliable broadband service available in our area, most of the people I knew were relying on their phones for Internet. Waiting around for my prescription to be filled, I grabbed my phone and found the site, which was nearly unusable with my phone. After zooming, swiping, and scrolling for about five minutes, I managed to highlight the first field to begin typing my name.

In this same period, in anticipation of the 2012 presidential race, political rhetoric about the "poor choices" of people still struggling to navigate postrecession labor markets had ramped up, and cell phones suddenly took center stage as the newest example of "wasteful spending" by the undeserving poor. Four years later, GOP hopefuls praised "gig" economy companies, like Uber, for creating employment opportunities for anyone with a phone.

In the past decade, the world's attention has been drawn to the ways that digital technologies have become central to "on-demand" forms of low-wage work. But I knew that, far beyond the rise of digital labor platforms, digital technologies had become infrastructural to new realities of precarious work across class divides. I knew that both white-collar professionals and low-wage service workers were relying on their digital devices to find work. But I had yet to figure out how they were both a part of the same story.

For three years, I interviewed both high- and low-wage precarious workers across the United States, as they patched together economic livelihoods. These workers used digital technologies in surprisingly similar ways: to find work, to maintain relationships, and to find dignity in often undignified conditions. In doing so, they used tech to solve the contradictions at the heart of capitalism. However, these practices were done in very different contexts, marginalizing some and privileging others, exacerbating an already polarized labor market. This book describes these practices and analyzes the hidden relationship between workers at both ends of our digital economy.

Acknowledgments

This book wouldn't have happened without the interviewees who were generous and patient enough to tell me their stories. Thank you to everyone who invited me into their lives for a few hours. I hope I got it right.

Writing this book took a more than a village; it took several. Through graduate school, a postdoc, and beyond, this manuscript has benefitted from generous mentors, colleagues, and institutions, as well as friends and family.

Thank you to the members of my dissertation committee at the University of Virginia. My chair, Sarah Corse, Allison Pugh, Krishan Kumar, and Siva Vaidhyanathan all encouraged me to be more careful with my thinking, be more ambitious with my claims, and stand up for my ideas within sociology and beyond.

When I was a graduate student, these ideas were improved by my fellow grads, especially Matthew Braswell, Francesca Tripodi, Claire Maiers, Benjamin Snyder, Christina Simko, Sarah Mosseri, and Catalina Vallejo, who all provided critical and supportive reads and hugs throughout the many stages of this project. The members of the Work/Culture Working Group, including Alison Gerber, Clayton Childress, Joe Klett, Ryan Hagan, Ryann Manning, Sorcha Brophy, and Ekedi Mpondo-Dinka, listened to my dissertation ideas as they morphed into something more.

Thanks to a host of early interlocutors as the dissertation started turning into a book. Late in my dissertation, I participated in a preconference put on by the Association of Internet Researchers' (AoIR) preconference on labor, where my words got a once-over from Elizabeth Wissinger, Airi Lampinen, Kylie Jarrett, Mary Gray, Karen Gregory, Dan Greene, Nancy Baym, and other participants. Around this same time, Jason Farman and Ben Snyder helped me figure out how to package the story as a proposal, and Jeff Lane and Sarah Brayne demystified the book process.

Thank you to the faculty and staff at The Institute for Advanced Studies in Culture (IASC) that supported me as a graduate fellow at the University of Virginia, especially James Davidson Hunter, Murray Milner, Katya Makarova, Garnette Cadogan, and Joe Davis. My time at IASC was brightened by the amazing Jeff Guhin, Tony Lin, Chad Wellmon, and Josh Yates.

This book and I benefitted from a postdoctoral fellowship at the Data & Society Research Institute. A huge thank-you to my postdoctoral mentor danah boyd, who gave me new shiny problems to dive into and allowed these ideas to simmer at a crucial time. Many thanks to Alex Rosenblat and Alexandra Mateescu, the women who showed me what true collaboration and intellectual kinship could really look like.

There were many others at Data & Society who alternatively held and pushed me along toward finishing this project. To Kadija Ferryman and Caroline Jack, my postdoc comrades who were such an important support through these amazing years. As the book became real, Angèle Christin, Luke Stark, Niels van Doorn, and the other participants at a 2017 Data & Society workshop helped me figure out what the digital hustle could be. Thank you to Andrew Selbst, Alexandra Mateescu, and Kinjal Dave, who marshalled the research on cell phones as essential infrastructure all the way to the Supreme Court! Thank you also to Markella Rutherford and Eni Mustafaraj at Wellesley College for their invitation to come back to present my work in 2017. And thank you to Lee Cuba and Tom Cushman at Wellesley for their early encouragement of my sociological imagination.

The transition to faculty life at the University of Pennsylvania brought new colleagues and collaborators. Thank you to Ben Shestakofsky and Lindsey Cameron, supportive colleagues who helped me find the puzzles that clarified my arguments. Thank you to my Annenberg colleagues, Barbie Zelizer, Joe Turow, and Jessa Lingel, for feedback on chapters and help with book-wrangling. A debt of gratitude to both Michael X Delli Carpini and John Jackson for their support, and also to Ryan Tsapatsaris and Helene Langlamet for their research assistance.

Thank you to Angèle Christin for organizing a manuscript workshop at the 2020 American Sociological Association meeting (and in all of our bedrooms and kitchens) and for being a touchstone for years through this project. A huge debt of gratitude for those who read and gave their feedback on the manuscript during a global pandemic: Matt Rafalow, Lilly Irani, Mary Gray, Alison Gerber, Taylor M. Cruz, Nick Couldry, and Clayton Childress. Many thanks to Caitlin Petre for her support through pandemic writing and parenting. Thank you to Cathy Hannabach and Candida Hadley with Ideas on Fire for their help with the finishing touches on the manuscript. And thanks to Dana Floberg for their policy prowess.

To James Cook, my editor at Oxford University Press, thank you for seeing the potential in this project at an early stage and for shepherding this project through the production process. Thanks also to the anonymous proposal and manuscript reviewers at Oxford for improving this project.

Last and certainly not least, I wouldn't be writing these words without the support of my family and friends. Thank you to my brother Adam Schroeder, Anne Rosenthal, John Updike, Lola Finocchiaro and Hollie MacLennan, Natalie Jo Ross, Alicia Skilinskis, Sara de Zarraga, Marta Daneshvar, Luis Ticona, and Justin Dick, who fed and watered me during periods of intense research and writing and who asked me how writing was going and listened while I complained. Thank you to my in-laws Clenith and Jose Ticona, who fueled my work with seco de carne and love. Many thanks to my own parents, Cheryl Martin-Schroeder and Mike Schroeder, who nurtured their bookish daughter to see herself not only as a reader but also a writer. They raised me to see that everybody who crossed my path had a story worth hearing. And thank you most of all to my son, Sam, whose pure joy for life lifts my soul, and to my partner, Dan, who has uprooted his life several times so that we could see where these ideas could go. Thank you for spending the first years of our relationship convincing me my ideas were good enough to write and the last few years of our marriage lovingly reminding me of all the good off the page as well. I love you.

Introduction

Precarious Inclusion in the Digital Economy

When I spotted her outside the Dunkin' Donuts in Washington, DC, where we agreed to meet for our interview, Jazmyne was glaring at her phone with a furrowed brow. When she noticed me, her face softened. "Sorry," she said. "They just changed my hours for tomorrow, and now I have to figure out who's going to watch my son." She made a quick call to her sister, and they compared their shifting schedules to figure out how to hand off her son. After a quick post online to check on the possibility of switching shifts, Jazmyne turned back to me and explained that she recently found a part-time seasonal job through Craigslist to supplement her hours stocking shelves at a big-box retailer. While she was excited for the extra cash, coordinating care for her son and an hour-long commute on unreliable public transportation were proving to be complicated.

She worked in one of the most expensive cities in the United States for slightly more than minimum wage and couldn't afford an apartment with enough room for her family, so, instead, Jazmyne took two trains and a bus to get from her jobs in the city to the apartment she shared with her sister in Maryland. She monitored accidents and delays in the city public transportation system from her phone throughout the day. If everything ran on time, she and her sister would get twenty minutes to debrief about her son's day before her sister would have to leave for her shift. As we took off our jackets and settled into high stools for our interview, Jazmyne set her phone face down on the counter in front of us, saying with a sigh, "I don't know what I would do without this thing. I even carry my charger around with me everywhere because it dies on me all the time and I'm so dependent on it! It's sad, but when my son's not with me, I'm checking it every five seconds. It's my crack. I'm totally addicted."

Jazmyne doesn't drive for Uber or work through an app, but for her, and many other low-wage workers in precarious work, smartphones and mobile access to the Internet have become essential infrastructure for their

Left to Our Own Devices. Julia Ticona, Oxford University Press. © Oxford University Press 2022.
DOI: 10.1093/oso/9780190691288.003.0001

livelihoods. Jazmyne's smartphone was a baton with which she conducted a chaotic symphony of daily insecurities, transforming them into a somewhat harmonious life for her and her family. Jazmyne enrolled her phone in strategies that allowed her to make do amid a shifting work schedule and complex negotiations of care.

Despite the increasingly important role digital technologies play in the lives of low-wage workers, the three major frameworks used to examine their role in the labor market have led to troubling gaps in our understanding. First, digital technologies are frequently misrecognized as relevant mostly to white-collar knowledge workers. The capacity to work from anywhere and blur the boundaries between the workplace and the rest of life are now no less important for precarious workers like Jazmyne than they are for highly paid knowledge workers. Second, when the technology use of low-wage workers is acknowledged, we often focus on "gig" workers who rely on apps and online platforms to find work. However, as this book explains, far beyond Uber or Amazon Mechanical Turk, digital technologies have become essential infrastructure for workers navigating precarious work across many different labor markets. Last, outside of the world of work, tech users like Jazmyne are often seen as being on the wrong side of the "digital divide," excluded from digital technology because of high costs or lacking the skills to use them. This framework misses that these workers are far from excluded. In fact, their digital inclusion has been weaponized by right-wing pundits and politicians who paint low-wage worker "choices" to invest in smartphones and data plans as wasteful spending. Each of these frameworks obscures the importance of digital technologies to workers like Jazmyne. These omissions have both analytical and political consequences as we continue to misapprehend the issues at the heart of our new digital ways of working.

Before the dramatic entrance of platforms into low-wage service labor, digital technologies like smartphones and laptops were largely considered the exclusive domain of white-collar knowledge work. These workers and their relationships with their technologies were seen as indicators of a new economic order.[1] From laptops that allowed them to work from virtually anywhere to the increasing expectation that they would work from everywhere, white-collar workers were at the center of the promise and perils of increasingly digital workplaces.[2] However, as work feels increasingly precarious for us all, the "always on" conveniences offered by mobile, Internet connected digital technologies are relevant for workers far beyond white-collar office work.

More recently, digital technologies have been at the center of debates about the rise of the so-called gig economy in the United States. Apps offering on-demand services like Uber, TaskRabbit, and Amazon Mechanical Turk have focused the public's attention on the ways some low-wage workers rely on digital technologies and Internet access for work. However, the workers who are heavily dependent on these apps to find work are a small proportion of American workers overall.[3] By contrast, workers like Jazmyne, who are precariously stitching together income from patchworks of jobs, are a much larger group of American workers.

There are substantial debates about what defines precarious workers; in this book, I define them as workers who don't have a single full-time job but, instead, regularly rely on income from multiple jobs, including part-time, contract, and informal work. These workers don't work consistent hours, their schedules often change at the last minute, and they have few benefits or guarantees of continued employment.[4] They work for low wages in retail or fast food, often combining formal and informal employment. Precarious workers can also be found in higher wage white-collar and professional jobs. Instead of being beholden to management or a single employer, their livelihood and professional identity depend on the maintenance of relationships that traverse the traditional boundaries of workplace and home.[5] Due to the conditions of insecure work, these workers are virtually required to maintain their connectivity simply to find and keep their jobs but are mostly absent from public debate about technology and the "future of work." As scholars and policymakers debate important issues surrounding the employment classification of platform-based workers, we're missing the fact that low-wage labor platforms are only the most visible tip of the iceberg of a much wider and less visible shift that has occurred in our expectations of universal connectivity for the purposes of work. Despite stagnating and insecure incomes, marginalized workers are purchasing and maintaining expensive smartphones and connections to the Internet because they're essential to re-shuffle their schedules, find childcare, or even just avoid boredom at tedious jobs, making it clear that the digital transformation of insecure work goes far beyond platforms.

We usually understand the technology practices of people like Jazmyne by looking at the ways they get excluded from accessing and using digital technologies. However, this focus on exclusion has hidden the unequal ways that economically marginalized people have included themselves, albeit within deeply unequal conditions. Pundits and politicians have used this inclusion to paint poor people's use of technologies like smartphones into morality tales about their individual responsibility for their own poverty. In response to a question about the high cost of health care for low-income Americans, former Republican Congressman Jason Chaffetz infamously suggested that people should simply make better "choices" about their spending, saying "rather than getting that new iPhone that they just love and want to go spend hundreds of dollars on that, maybe they should invest in their own health care."[6] Just as big-screen televisions and cable subscriptions were once used as examples of irresponsible consumption, smartphones have become a symbol of wasteful spending, individualizing the social problem of poverty. In focusing on exclusion, we've missed the ways social inequalities shape the digital inclusion of different groups, producing patterns that marginalize some and privilege others.

To understand digital technology's role in insecure work, I took a comparative approach, interviewing one hundred high- and low-wage contingent and independent workers in four different cities across the United States. I hung out in the big-box stores, neighborhood sub shops, and luxury retailers where people buy their phones and talked with them about how they used all kinds of digital technologies to navigate their work and the rest of their lives. Based on these interviews, I argue that Internet-connected devices have become infrastructural to insecure ways of working, meaning that they have sunk into the bedrock of taken-for-granted tools for daily life.[7] Faced with unpredictable jobs and an unraveling social safety net, independent and contingent workers from across the income spectrum depend on their digital technologies for economic survival and to construct identity across workplaces, clients, and gigs. However, despite these similarities across social class, I find that the terms on which these groups are included into digitally saturated labor markets reproduce, rather than reduce, inequalities between classes.

In our haste to chase the "future of work" with new technologies, we have neglected to fully understand its present. Ignoring the important role of these technologies in our economy has political consequences. As Lisa Parks observes, "When technologies remain hidden or obscure, they remain beyond public concern."[8] Allowing this infrastructure to remain invisible isn't only a problem for scholars; it's also a problem for policymakers and others who'd like to work toward more equitable connectivity because it forecloses many paths toward intervention and reform.

Invisible Infrastructures of Insecure Work

For about a decade now, the digital technologies that facilitate insecure and nonstandard work have become closely associated with the idea of the gig economy and a handful of apps and platforms that seem to be at the center of it. As the story goes, through their technologies, companies like Uber, Lyft, TaskRabbit, and Amazon Mechanical Turk have spawned a new kind of relationship between workers and their employers. These apps and platforms have been at the center of public debates about the "future of work," both celebrated as ushering in the dawn of a new liberating era of capitalism and excoriated as exploitative mechanisms of neoliberal control. This debate has raised important questions about technology's role in the administration and culture of capitalism. However, these headlines have hidden a quieter but more profound shift in the relationship between insecure work and digital technologies. Digital devices like smartphones and laptops have become deeply embedded into the infrastructure of insecure work far beyond the handful of apps we read about on the front pages. Although they no longer grab headlines, these technologies have been quietly facilitating the work of both high- and low-wage contingent workers in many different types of jobs across the economy.

The gig economy entered the American lexicon in the aftermath of the 2009 recession. As a severely contracted labor market saw massive layoffs and few new jobs, websites and services often referred to as the "sharing economy" promised to help Americans share resources or make money by renting their idle cars, tools, and rooms. Amid a glut of unemployed and underemployed college graduates and many other highly qualified workers, platforms such as Uber and TaskRabbit offered ways for people to earn cash doing short-lived tasks for others.[9] Workers use these apps to find work as independent contractors, and the companies that hire them have very few obligations toward them.

While on-demand apps have sparked a broader public discussion about the issues faced by these workers, on-demand apps didn't invent this kind of work. Since the mid-1970s, workers in both high- and low-wage work have dealt with the "fissuring" of work, or the dismantling of traditional forms of employment and the rise of contingent, temporary, and precarious or risky forms of work.[10] This huge shift in the obligations employers have toward their employees has brought increased flexibility in day-to-day scheduling, casualization in job tenure, and the individualization of risk and responsibility for job success.[11] These changes have shifted many of the risks of employment onto workers, as health insurance and pensions become workers' problems instead of employers' problems.[12]

If we look at gig work from this broader perspective, as insecure work rather than only those jobs found through a platform, it's clear that most gig work doesn't actually take place on platforms. Estimates put the number of Americans earning income from platforms at about 1 percent, while estimates of workers in precarious or contingent work range between 10 and 40 percent.[13] This range makes it clear that definitions of this type of work vary widely and that it's proven difficult to measure the ways Americans are stringing together multiple jobs at once and over time. However, despite some disagreement over their precise numbers, the online platforms that account for a much smaller proportion of workers have taken center stage in our public conversations about digital technologies and work, while the technologies that scaffold the experiences of a much larger population of workers have received much less attention. This has led to vigorous debates about the legal employment status of platform workers, and importantly, it has left little room to debate the now central role of digital technologies in scaffolding precarious work much more broadly.

On and off the job, our lives are increasingly facilitated by our ability to connect to the Internet and one another through digital technologies. However, the rate at which households have subscribed to home broadband service and use laptops and desktops to go online has slowed in recent years.[14] Meanwhile, both cell phone and smartphone ownership have increased tremendously. For more than one in six Americans, smartphones are their only access to the Internet.[15] In fact, mobile phones aren't only people's way of accessing the Internet; they're increasingly their only phone. More than half of US households have a cell phone but no landline.[16] Recognizing the changing role of these technologies in our lives, in 2018 the Supreme Court ruled that cell phone location information should be afforded a higher standard for privacy because, in large part, cell phones are increasingly mandatory for everything from public safety to accessing medical information, social services, or even looking for a job.[17]

Despite their ubiquity, social inequalities pattern Americans' use of digital technologies to access the Internet. Households where cell phones are the primary Internet access and telephone aren't equally represented across society; they're concentrated among young, low-income, and Black and

Hispanic communities.[18] People who are poor and working class, as well as members of Black and Hispanic communities, also rely more heavily on their phones to accomplish more complex tasks, like looking for jobs, filling out online job applications, and creating resumes.[19] These same populations are also more likely to find themselves in precarious jobs, creating a perfect storm of smartphone dependence and economic insecurity.

How is it possible that these technologies are both ubiquitous in our daily experiences and also largely absent from our public conversations about technology, work, and inequality? This book looks at these apps and websites alongside other hardware and software that are less visible, but more ubiquitous, technologies of work—including cell phones, laptops, social media sites, calendar apps, and text messaging platforms. These technologies aren't at the bleeding edge of new media or technological disruption anymore, but their ubiquity is what makes them even more important to understand.[20] Talking with workers across the economy made clear that these technologies, once the center of breathlessly anticipated announcements, are more important than ever precisely because they've faded into the background or infrastructure of our taken-for-granted, everyday habits of working.

Inclusion and Inequality

Stories like Jazmyne's are usually told in the context of the "digital divide," a term that refers to a gap in access to technologies, skills, and habits cultivated by wealthy and poor users. The digital divide helps us conceptualize the ways well-resourced workers have access to the newest and fastest technologies as well as robust and nearly ubiquitous Internet access and possess relatively sophisticated abilities to use these technologies, while workers like Jazmyne struggle to put together the funds to purchase and maintain pricey devices and ensure continuous Internet access.[21] But, as this book will explain, low-wage workers aren't being excluded from using digital technologies or the Internet; quite the opposite, their work increasingly requires it. Jazmyne is neither disconnected nor unskilled, but our current frameworks make it hard to see how social inequalities shape her experiences. As this book argues, it's not only exclusion from digital technologies that structures inequalities in opportunities but inclusion as well.

Digital inequalities are primarily considered a problem of exclusion from the benefits of digital technologies, but it's important to also understand the unequal terms on which people are included in the digital economy. This shift to inclusion allows us to ask questions about the ways different groups use digital technologies in the context of already existing sets of social relations characterized by different amounts of power and privilege. Thinking about the terms of inclusion allows us to examine how the technology use of different social groups is received or understood by outsiders. [22] As I'll explain, the digital inclusion of working-class people has been weaponized by right-wing politics. Thinking about inclusion also opens us up to questions about the role of privilege in technology use. Digital technologies are becoming important players in processes of social stratification, and these processes reproduce both marginalization and privilege. By focusing exclusively on those who've been marginalized by our current digital arrangements, we've neglected to examine the people who've benefited from them. Studying the terms of inclusion shifts our attention away from the things that low-wage workers supposedly lack and toward the social relations that condition their use in ways that reinforce, rather than alleviate, social inequalities.

The "digital divide" model began its life as a policy framework animated by the assumption that exclusion is the root of digital inequality. This model contributed to a narrative about the supposed democratizing power of digital technologies to ensure labor market opportunities. In the 1990s, globalization and deindustrialization saw the decline of US manufacturing and the chipping away of organized labor, causing structural disruptions to working-class communities, particularly Black enclaves in urban centers like Chicago.[23] Ensuing political debates about how to address the complex sources of poverty produced a discourse about the success of the burgeoning industry of information and communication technologies. This "access doctrine" provided the common sense that individually acquired digital skills, not structural intervention, would be the answer for workers displaced by these structural shifts in the US economy. Digital technologies became the center of many US efforts to ensure labor market opportunities.[24]

The digital divide model has been persistent in our politics and has also shaped social science research.[25] In studying these issues, scholars first focused on the ways that social inequalities shape who gets excluded from the access to computers and the Internet.[26] They then turned to examining the ways some groups are excluded from developing necessary digital skills or from using technology to develop their human capital. Researchers also examined the ways social groups differed in their orientations toward technology and the Internet, leading some to be excluded from the benefits of digital technologies and the Internet. These exclusions aren't random but are socially patterned in ways that reinforce existing forms of social stratification.[27] As Castells explains, "exclusion from these [Internet] networks is one of the most damaging forms of exclusion in our economy and in our culture."[28] Whether looking at access, skills, or motivations, we've mostly understood digital inequality as a problem of exclusion.

As scholars focused on the exclusion of marginalized groups, right-wing politicians were busy weaponizing the digital inclusion of poor and working-class people. From Jason Chaffetz's dismissive comments about low-income people needing to choose between their iPhone and health insurance to efforts to limit the Lifeline program, which provides modest subsides for Internet subscriptions and mobile phones, marginalized people face skepticism from the public about the "necessity" of sophisticated smartphones to their everyday lives. From the beginning, it was clear that class has been shaping the moral discourse about digital technologies. In 2001, Federal Communications Commission (FCC) Commissioner Michael Powell commented on the "so-called digital divide," saying that "I think there is a Mercedes divide. . . . I'd like to have one; I can't afford one."[29]

In 2005, George W. Bush's FCC expanded the Lifeline program (originally established by President Ronald Reagan at the request of a bipartisan group of legislators and which connects rural and low-income Americans to telephone services) to include subsidies for mobile phones. In 2016, President Barack Obama expanded the program again to include broadband subscriptions.[30] However, as programs like Lifeline and the falling cost of smartphones and mobile Web access have begun to close stubborn divides in ownership between rich and poor, and between Whites and Blacks and Latinos, a growing cultural backlash has pulled digital technologies into a vocabulary that demonizes marginalized groups for their wasteful spending choices.

Lifeline's mobile phone program, derisively and inaccurately called "Obamaphones," became a flashpoint in conservative circles in the run-up to the 2012 presidential election between Barack Obama and Mitt Romney. In 2011, Tim Griffin, a Republican congressman from Arkansas, sponsored the Stop Taxpayer Funded Cell Phones Act that would've prevented mobile carriers from receiving payments from the fund that reimburses them for subsidized Lifeline subscriptions, effectively shuttering the program's efforts to provide mobile services. In 2012, a few months after Romney's infamous and false comments about the "47 percent" of Americans who are dependent on government handouts, Lifeline phones went viral in a YouTube video depicting a woman who became known as the "Obamaphone lady." This interview with an "Obama voter" featured a Black woman attending a protest outside a Romney rally in Cleveland, Ohio, saying she was voting for President Obama because he gave her a free phone. This video was snapped up by conservative media pundits, discussed several times on Rush Limbaugh's radio show, featured at the top of the Drudge Report, and widely discussed on the White nationalist site Stormfront.[31] The "Obamaphone lady" video was traded among conservative pundits as the "welfare queen" trope of the digital era, gleefully described in coded language that reinforced racist and gendered stereotypes of those receiving public assistance.[32]

These stereotypes showed up in some of my higher wage interviewees' language for talking about the technology use of the lower income people they came into contact with throughout the course of their workdays. Lucy, a nurse in a practice that served a low-income community in rural Mainville, admitted:

> I may be a little jaded because of the work that I do; there's a certain population that you see just seem to not care about any of that stuff, about their privacy.
>
> Julia: What do you mean?
>
> Hmmm. Well, let's see. It's people that abuse the health care system because they don't want to take responsibility for their own actions. They feel crappy, but they're 100 pounds overweight, and they smoke, and they drink, and they eat a terrible diet, and it's always, "My husband left me, and I can't get the disability that I think I deserve" and "Will Medicaid cover this?" . . . It's the population that I work with . . . but I find it a little discouraging. And those are the people, every one of them has a smartphone. It rings, and they answer it, even in the middle of an exam! It's like, how can you afford that? My bill is always so high!

Smartphones have become a symbol in right-wing political culture, crystallizing feelings of resentment from working-class White people and their wealthier sympathizers against Black and brown people they see as the undeserving poor. While social scientists have argued that inclusion is essential to solving inequalities of opportunity, from this perspective, digital technologies are irrelevant to the lives of marginalized populations or, even worse, emblematic of greed, entitlement, and poor self-discipline. Such a perspective erases generations of structural inequalities and institutional discrimination.[33] These politics acknowledge the digital inclusion of lower income people but have turned it into an argument to support gutting the meager supports that have enabled that inclusion in the first place.

The ways we, as individuals, use digital technologies are enmeshed in existing relationships of power between groups. The digital inclusion of poor people has been enrolled in projects that marginalize their use and also uphold the practices of middle-class professionals as correct and upstanding. Examining the ways inclusion leads to inequality also attunes us to the role of digital technologies in social processes that benefit some social groups over others. Digital technologies first became ubiquitous in the world of white-collar work and were largely studied within these contexts, where workers enjoyed livable and largely stable incomes, predictable schedules, and the autonomy afforded to professional workers. For these privileged workers, stable labor market conditions were layered on top of unproblematic and nearly universal Internet access at home and at work, updated and sophisticated software and hardware, and organizational acknowledgment (and sometimes even financial support) for the necessity of connectivity for work. As a result, the early and remarkably well-supported digital inclusion of white-collar workers has become the unmarked standard by which we measure marginalization.

As this book will describe, digital technologies, especially smartphones, are increasingly ubiquitous up and down the income ladder. Workers are using their devices to navigate informal, black, and gray labor markets, in workplaces that share very little with white-collar contexts. As work feels increasingly precarious for many workers, it is necessary to recognize how the conditions we often consider "normal" are actually contingent on workers' privileged position within existing social relations. This is an especially important shift to make when fewer and fewer people fit into that mold and are finding themselves in new kinds of employment relationships.

The Meaning of Inclusion

Digital inequalities aren't only visible in the kinds of devices people have or how they use them but also in how technologies fit into the stories we tell about ourselves, from our values to the ways we see ourselves as workers. These inequalities are evident in Jazmyne's feelings of helpless dependence.

Digital technologies are often seen as alienating workers from meaningful work. We're told that we're addicted, dependent, or even tricked by design to seek out constant connectivity at the expense of authentic and meaningful interactions with colleagues or focused productivity. However, as this book will argue, how insecure workers feel about their digital work practices actually shows the important role they play in the construction of identity and dignity in these types of work.

Economic precarity isn't only an economic phenomenon but also an affective one. Precariousness describes a particular set of economic circumstances and is also marked by a sense of anxiety and insecurity. Our expectations and experiences of economic insecurity carve deep grooves into our feelings and identities.[34] From partnership and romantic relationships to our expectations of our careers, structural economic shifts don't only affect our economic prospects but also the stories we tell about ourselves.

Structural changes, like the shifting of risks toward workers and the demise of the "standard employment relationship," have engendered shifts in relationships with employers as well as cultural shifts in workplaces and workers' experiences. Higher wage workers have been both pushed into independent work by these longer-term shifts and pulled by new kinds of ideologies about work, valuing entrepreneurialism, risk-taking, and individual reputation over things like loyalty, seniority, and stability.[35] As formal organizations and institutions that govern work distance themselves from workers, or fade from the foreground of workers' daily experiences, these ideologies and narratives become central to building identity.[36]

These ideologies are powerful, allowing workers to transform precarity into a meaningful and positive identity around work. Studying workers in the midst of the heady dot-com bubble, Gina Neff observed the powerful pull of this ideology as it encouraged even salaried workers to think and act like entrepreneurs, performing "venture labor" or embracing risks on behalf of their companies, even without any real ownership stake. These new ideas about work suggested that risky work wasn't just what was happening to your job; it was also desirable and cool.[37] In an economic environment that provides fewer paths to security and less control over labor market outcomes, risk-taking provides workers a feeling of choice, a seductive, but likely hollow, offer to take control into their own hands.

Long before apps and smartphones, digital technologies became associated with more precarious and short-term employment work relationships and linked with ideas about the freedom of individual entrepreneurship.

As computers evolved from room-sized hulks to desktop machines and the Internet emerged onto the public stage, cyberutopians linked their ideas about the promises of these developments with emerging forms of work organization.[38] By the mid-1990s, thinkers like Nicholas Negroponte explicitly linked digital technologies to flatter organizations, more democratic employee governance, and more freedom for workers as they moved across a flexible economy.[39]

Later on, digital technologies took on a less utopian cast. As digital technologies settled into mundane tools in white-collar workplaces, scholars observed the synergistic ways they seemed to amplify the effects of economic insecurity. Increasingly, mobile digital technologies allowed workers to use them in ways that accelerated workloads and disembedded workers from the formerly separated spaces and times of work. In her study of Australian workers on the margins of white-collar professional work, Melissa Gregg found that the practices of remote work, facilitated by laptops, mobile phones, and home Internet access, heightened workers' fear and sense of threat.[40] Over the past several decades, concerns about rising expectations of constant connectivity and overwork, or the health risks of sedentary work in front of bright computer screens, have galvanized support for reforms surrounding "work-life balance"—such as France's decree against after-work emails and a cottage industry of self-help books and software designed to impose boundaries between work and home life. As white-collar workers settled into precarious working conditions as their new normal, the ways digital technologies were changing white-collar work practices were seen as amplifying risk shifts and emptying work of meaning.

Accelerated by the dot-com boom, the desirability of risky work is a trend we generally associate with white-collar knowledge workers, but boosters of gig economy platforms like Uber and Handy stitched associations between digital technologies and freedom into the world of low-wage service work. As explained earlier, low-wage workers have been coping with precarity on a much longer timeline than higher wage white-collar workers. However, in the early years of the twenty-first millennium, the leaders and other boosters of labor platforms such as Uber, Lyft, and Handy built a narrative about the power of digital technologies to put low-wage workers in charge. Oisin Hanrahan, the CEO of Handy, a housecleaning and home services platform, explained that his app was "really changing what was a very regimented and structured way of working, where the company . . . essentially had all the control . . . to one where you really are flipping the entire thing on its head and saying . . . you can decide wherever you want to work. . . . There's a huge opportunity here to use technology to actually empower workers."[41] These leaders transposed early utopian ideas about computers and white-collar freedom into the practices of low-wage service work, reinvigorating the symbolic associations between digital technologies, freedom, and work.

While the promises of putting control back in the hands of low-wage workers were far from reality, these associations between digital technologies and more autonomous and dignified work remained. As tools are central to the execution of our work today, our practices with digital technologies still play a constructive role in building and maintaining meaningful work identities. From taking pride in promptly answering the pings and pulls of emails or texts from colleagues to deftly soliciting a glowing online review, my interviewees' skilled and practiced use of the digital technologies of their work played a large role in the ways precarious workers constructed their identities and proved to themselves and others that they have what it takes to make it in cutthroat labor markets.

Scholars of craftwork point to the importance of work practices in identity creation. Craftworkers, from woodworkers to potters, develop their skills through continuous practice with the tools of their trades. Over time, they develop "virtuoso" skills with those tools and materials, which include mastering many different techniques and doing things with speed and agility.[42] These work practices bind craftworkers into moral orders of worthiness and are essential ingredients in understanding the pride and dignity workers take in their work.[43] Indeed, the skilled use of tools in the pursuit of "good work" may be the craftworker's "primordial mark of identity."[44]

The study of digital technologies like smartphones and laptops may not seem like a natural fit with ideas about craftwork. Indeed, scholars have linked the increasing use of technology in the workplace with the erosion of craft practices by capitalism and globalization.[45] However, I found that workers' practices with their digital technologies were a source of pride, leading me to interpret them as a new kind of craft labor for insecure work. Working as they do between different organizations, clients, and types of work, precarious workers are often without the traditional indicators of a job well done—things like performance reviews, raises, or other kinds of "gold stars" that indicate good work. As a result, independent and contingent workers construct their own proof of their worthiness and look for signs of validation that they're good at their jobs.

Over the past decade, many have attributed new kinds of emotional experiences and feelings to the use of an ever-expanding ecology of digital technologies, apps, sites, and social media platforms. Whether tricked by addictive features designed to manipulate our brains, seduced by empty promises of intimacy and social connection, or pressured by toxic organizations, we find our digital devices to be a source of both consolation and frustration. But it would be a mistake to try to understand how our feelings are entwined with our digital devices and our work without understanding their connection to an insecure and polarized labor market. The embedding of digital technologies into precarious work allows us to see how our digital technology practices can also be tightly linked with our identities, dignity, and sense of self-worth in uncertain economic times.

The Study

Despite the spread of precarious and contingent work across social classes, we don't have many studies that compare the experiences of high- and low-wage workers doing this work, especially around the issue of digital technology. The labor market prospects of high- and low-wage workers have sharply diverged since the 1980s. Since this time, the United States has witnessed an uninterrupted growth in economic inequality, hollowing out the middle class and driving a larger wedge between those at the top and those at the bottom of the income ladder.[46] Studies about the role of digital technologies at work have mostly followed suit, focusing either on white-collar freelancers or low-wage precarious workers.

Studies that have examined the role of digital technology at work for high- and low-wage workers separately have focused on the role of different types of technologies for these two different groups. For high-wage white-collar workers, the tools of digital knowledge work, like laptops, the Internet, and smartphones, have enabled more freedom in the time and place of work but have also fostered burnout as work creeps into other parts of life. For low-wage precarious workers, on-demand platforms—such as Uber, Lyft, and TaskRabbit that allow them to offer in-person services—and automated application or scheduling platforms exert a cruel and exploitative algorithmic control over workers without the oversight of human managers. Together, these studies paint a picture of digital technology use that is very different for workers of different social classes. But, with the spread of contingent working conditions across classes, the question remains: How do the conditions of contingent and precarious work shape workers' relationships with the digital technologies that are increasingly required for their work?

To avoid sampling on the dependent variable, instead of choosing interviewees by the type of technology they were using, I recruited them based on the types of work they were doing. This book is based on ethnographic interviews with one hundred high- and low-wage precarious workers in four different cities across the United States. I first met the people who would become interviewees for this study in the places they shopped for and purchased their digital technologies: consumer retail stores. In each city, I used publicly available demographic information and the advice of local guides to choose neighborhoods where high- and low-wage workers might live and scouted those neighborhoods for the retail stores that sold digital information and communication technologies to consumers. Traditionally, qualitative studies of work focus on workplaces or other physical locations where the workers scholars want to study gather together. However, the contingent and independent workers I was interested in talking to often don't have a single workplace but are instead constantly moving among clients, employers, and even types of gigs, sometimes all in the same day. For these reasons, I chose to focus on recruiting in retail locations where these workers purchase and service their digital technologies.[47]

Each interviewee picked the place where we conducted the interview. I interviewed people at their dinner tables and in coffee shops, parks, and libraries across the country. During interviews, I asked workers to tell me stories about how they use digital technologies throughout their daily work and to navigate their lives in nonstandard work. I was interested in their answers and in the tradition of ethnographic interviewing. I also paid attention to how they told their stories and interacted in the settings they had invited me into. To better understand the contexts of digital technology use for high- and low-wage workers, I also spent time in the fast-food joints and subway stations where many of my interviewees accessed free Wi-Fi and talked to some of the "brokers" of their access, such as cell phone salespeople, gas station cashiers, and store managers.

Throughout my fieldwork, I came to realize that consumer electronics retail is highly segregated by class. I found that, even in my rural fieldsite, high-wage workers were unaware that a local sub shop and gas station sold phones while low-wage workers may have browsed but had never purchased anything from the Apple store at a nearby mall where many of the high-wage workers regularly bought their technologies. Major mobile phone carriers oftentimes even segment their brands that cater to low-wage consumers to avoid association with their main brand identity, which is usually aimed at higher income customers.[48]

In this book, I pay the most attention to the ways social class, as it's constituted by high- and low-wage work, structures workers' everyday experiences with digital technologies and the Internet. In studying work and workers, social class is often taken as the social classification most consequential for understanding inequality. This analysis follows in this tradition. However, class is only one axis in the matrix of classifications—including race, gender, and sexuality, and many others that pattern workers' experiences of and ability to exercise power.[49] These categories are socially constructed forms of classification that have real power to shape social realities; they are also cultural frames that shape how the technology use of different workers is perceived and interpreted by others.[50] However, all of these categories are not equally salient all the time.[51] To this end, throughout the chapters, I also consider the important roles gender, race, and sexuality play in workers' use of their devices. More detailed information on the research design, sampling, and recruitment of interviewees can be found in the methodological appendix.

The Argument

Based on this data, *Left to Our Own Devices* tells a story about surprising similarities in work practices across classes but also of the vast differences in the contexts where workers set them up and put them to use, leading to a deepening inequality between high- and low-wage workers. Despite differences between these groups, I find that digital technologies are essential tools for both high- and low-wage contingent workers. Both sets of workers constructed a similar set of strategies I call the "digital hustle" to get, maintain, and carry out their precarious work. The broad use of the digital hustle across classes illuminates how these technologies have become infrastructural to facilitating current regimes of independent and precarious work in the United States. These strategies are widespread because they're "biographical solutions to systemic contradictions," meaning they allow individuals to cope with and adapt to structural economic changes, like the deregulation of markets, the pressures of global competition, and technological changes in employment, over which they have little agency.[52]

These strategies help both high- and low-wage workers piece together patchworks of paid gigs across many different labor markets. But they aren't only oriented toward the market; they are also oriented toward the self. The digital hustle is tightly linked with workers' identities and their sense of dignity in insecure circumstances. In work that takes place outside of and across different organizations and teams, practices of promptly answering emails or texts from clients weren't only economically important, they were also indicators to workers that they were good at their jobs. These practices are often seen as signs of a pathological relationship between workers and their work, but I point out how they're also strategic performances of moral worthiness for an audience of one, as workers execute these strategies to gain a sense of autonomy in contexts that are often hostile to it. These technologies, and the strategies workers construct with them, allow workers to accommodate themselves and their identities to insecure work, to bend so they do not break.

I call these sets of work practices "strategies" because they draw on the tools and resources workers have "ready at hand."[53] Instead of understanding such activities as rationally planned or preselected, theorists who emphasize the importance of "practice" point out that the ideas or actions people take in the moment are cobbled together from the resources available in the environment. I understand these resources as consisting of the digital technologies, Internet access, and cultural resources like authority and autonomy afforded in their workplaces.[54]

Although these workers had similar strategies, the digital hustle ultimately reproduces inequality between high- and low-wage workers in insecure labor markets. As I explain in Chapters 2 and 3, high- and low-wage workers are included in the digital economy on terms marked by stark differences in power and privilege. First, the political economy of consumer markets for digital technologies privileges the already advantaged and exploits groups who've faced institutional discrimination from banks and other financial institutions. Second, workplaces, managers, and clients recognize the savvy and skilled use of digital technologies as essential for high-wage workers but don't recognize, and sometimes even actively penalize, these same uses for low-wage workers. I argue that the sharply different conditions under which these types of workers are included into the digital economy mean that the digital hustle reproduces rather than equalizes inequalities in insecure labor markets.

Overall, these findings show how digital technologies have become the infrastructure for workers' ways of coping with insecure work across high- and low-wage labor markets. These strategies are a coping mechanism for economic survival, and they're also central to the construction of identity for workers who labor at the margins. Despite widespread assumptions about the role of digital technologies in equalizing entrepreneurship and economic opportunity, I show how the intertwining of precarious and contingent work with digital technologies reproduces and exacerbates existing social inequalities between high- and low-wage workers. This book illuminates that digital inclusion is not enough to secure equality and may in fact deepen the divides between us. Despite the widespread use of digital technologies across social classes, *Left to Our Own Devices* shows how the unequal and exploitative conditions under which workers are included in the digital economy can exacerbate existing inequalities.

Organization of the Book

This book is organized to reflect the "doings" and "feelings" of high- and low-wage precarious workers. In Chapter 1, I describe the digital hustle, a shared set of coping strategies that both high- and low-wage workers used to find and coordinate work. While offering handyman services on Craigslist and maintaining a LinkedIn page may not appear similar at first, I illustrate the hidden convergences behind these disparate practices. The digital hustle empowers workers to piece together patchworks of paid work in the margins of high- and low-wage labor markets, but because the digital hustle is unpaid, it may exacerbate inequalities between workers. The digital hustle isn't only used to meet economic needs; it also plays an important role in the constitution of identity and provides a source of dignity within often humbling labor market conditions.

While the digital hustle is a shared set of strategies, high- and low-wage workers don't do it under the same conditions. In Chapter 2, I explain how the terms on which low-wage workers are included in the digital economy reinforce and amplify the disadvantages they already face in the labor market and elsewhere. From buying phones at Walmart and buying airtime at gas stations, low-wage workers face an exploitative marketplace to purchase phones and Internet connections. In addition, these workers often utilize technical infrastructure—from public or commercial Wi-Fi networks to phones with poorer privacy protections—that makes them vulnerable to increased amounts and invasive forms of data collection. Market actors make connectivity appear affordable, but predictable instability in their connections requires workers to do "invisible labor" to overcome being disconnected. I also describe the crosscutting ways that racism and the threat of policing and violence shape workers' ability to use Wi-Fi in public places.

In Chapter 3, I turn my attention toward the high-wage workers. While low-wage workers are often understood as scrambling over hurdles that interfere with their ability to use digital technologies in the same ways as high-wage workers, in this chapter, I flip the script to understand how high-wage workers are the beneficiaries of "digital privilege" that facilitates their technology use for work. I illustrate how ubiquitous connectivity is naturalized, expected, assumed, and rewarded, and how it is also facilitated by their capital and workplace contexts.

Chapter 4 compares how both high- and low-wage workers both use and refuse to use their digital technologies in strategies to resist work. While both groups of workers talked about digital technologies in their efforts to resist work and carve out spaces of autonomy for themselves, they did so in very different ways depending on the forms of managerial control they faced. Low-wage workers resisted direct control over their bodies and attention by diving into their phones to create a kind of absent presence. By contrast, high-wage workers refused to use their digital technologies to limit their affective commitment to jobs that demanded their devotion. However, I also explore how these strategies aren't available to all workers in the same ways. This chapter explains how race, sexuality, and gender constrain workers' strategies to use their digital technologies to resist work.

In the Conclusion, I reflect on the political consequences of leaving workers to their own devices as well as on the consequences of thinking about inclusion as an important facet of digital inequality. I explain the implications of the book's findings for researchers concerned with digital inequalities as well as for people working to solve some of these pressing issues. The consequences of our deepening dependence and expanding ecology of digital technologies in the world of work shouldn't only be a concern for Uber drivers and factory workers whose jobs are under threat from automation; it should be a concern for all of us. As we stare down the "future of work" for US workers on either side of a polarized labor market, the mistaken belief that digital technologies democratize access to the American dream of economic mobility through work affects us all and has serious consequences for our ability to understand our new ways of working.

1

The Digital Hustle

When we met in the middle of a rare snowstorm in Washington, DC, in January, Charlie was bundled up against the cold in his Carhartt jacket, thick socks, and sturdy work boots, with a knit cap pulled down over his ears. As he peeled off his many layers in our booth at a Dunkin' Donuts, he apologized for smelling like cigarette smoke, saying that bad winter weather always makes him think a little harder about quitting for good. Charlie explained that smoking was a small comfort in what he felt were uncertain times. "It's like, every day you just you walk out your door and you're already stressed. Because we never know, even these days, you never know what the next day is going to be like. You have no idea. I'm just trying to keep my guys busy."

Charlie's "guys" were a small crew of two or three manual workers he tried to keep in regular work through a patchwork of contracting, demolition gigs, and moving jobs. Looking older than his forty-seven years, Charlie told me about how he came to start his own home contracting and moving business after he left his union construction job when his boss was replaced by someone much younger than he was. He enjoyed the freedom and independence that came with "being his own boss":

> Being my own boss, I don't have to deal with nobody. And for me, because I'm forty-seven, I can't deal with a twenty- or thirty-year-old, some young kid like you being my boss.
> Julia: [laughs] Why not?
> Just can't. It just gets under my skin, I mean right? You're young, maybe you'll see someday.

While he liked the freedom of owning his own business, Charlie spent much of his time and energy working to secure future gigs. This work involved constantly posting and monitoring Craigslist and other platforms that facilitate gig or short-term contract work from his smartphone and laptop. As he explained, this required a lot of time and energy spent on his phone:

Left to Our Own Devices. Julia Ticona, Oxford University Press. © Oxford University Press 2022.
DOI: 10.1093/oso/9780190691288.003.0002

And I always answer my phone, even when I don't recognize the number because people are calling you through Craigslist or they got my number somehow and want me to work. . . . Just like when I was coming over here, I actually was on Craigslist posting my ads on it. . . . I post my ads but I have to keep an eye on them because they disappear after other people post new ones, so I have to make sure people continuously see me. A couple of weeks ago I went San Diego [for a vacation]. I brought my laptop and my cell phone . . . it was as if I was over here [at home] . . . just keeping an eye on my work that's going on. . . . Calling [my workers] on their cell phones, doing work off Craigslist.

Charlie gets many of his gigs from ads he posts on Craigslist, which requires him to stay vigilant about answering his phone, posting his ads, and coordinating his workers. Being visible in online spaces and attentive to messages and emails was vital to Charlie's business, ensuring that he can "keep his guys busy" with a steady stream of work.

While Charlie's practices were motivated by the need to sustain new business, money wasn't the only thing that drove his daily routines. The frenetic and constant use of his digital devices was central to how he constructed his identity as a worker in his field. To find moving gigs, Charlie used an online marketplace that connects businesses with shippers with extra space in a truck or who run their own small trucking or moving operations to save on traditional shipping companies. This marketplace, as well as many other "sharing" or "on-demand" economy platforms, relies on customers to rate shippers through a five-star rating system. Shippers bid on jobs through an online auction, and customers use shippers' profiles, which display detailed ratings and comments from previous customers, to assess and choose a provider. While for most of our interview Charlie narrated his work with resigned pragmatism about the state of the economy and the future of his business, when he talked about working as a mover through this online marketplace, a small measure of pride crept into his voice as he told me:

It doesn't even pay as good as construction, but I've been a mover for more than twenty years, and I'm a really good one. [smiles]

Julia: How do you know you're a good mover?

I'm always checking my stars and asking people to rate me. I got a pretty good rating on there, too. I'm a damn good mover [smiles], and people need to see that, there's a lot of bad people out there trying to scam people in moving. . . . I grew up in the '80s, so that's a hard-working generation. I try and teach the young one who works for me about all this stuff . . . making sure you answer people right away, make sure you're always giving them fair prices, that's what gets you good ratings and brings in more business. He's grown up a lot. You have to learn from your elders, because there's nobody else going to teach you anything, I expect no less from him than I do from my own self. He knows that.

Charlie's ability to maintain his online reputation in this marketplace was important to securing future business. Five-star rating systems have become ubiquitous in many parts of the service economy, as customers are prompted to quantify their satisfaction with their experiences everywhere from hotels to doctor's offices.[1] These rankings are a form of discipline for workers, as they serve to classify, standardize, and hold them accountable to oftentimes impossibly high standards and capricious tastes of customers.[2] Workers and customers, in turn, react to ratings systems in ways that often redefine the values they purport to measure, adjusting expectations and performances to fit the measurement.[3] However, for Charlie, those five stars indicated more to him than just his likelihood of gaining more customers; they were constitutive of the way he understood himself as being good at his job. Charlie links his "pretty good rating" to being honest, forthright, and fair to his customers as well as to the importance of socializing one of his young workers to these same "hard-working" values. While Charlie's rating is oriented toward the market in that it's important to maintaining his client base, it's clear in Charlie's comments that it's also oriented toward the self, reinforcing his identity as someone who upholds certain values in his search for independent work.

A fifteen-minute bus ride away from the Dunkin' Donuts where I met Charlie, I rode the elevator to a stylish open-concept office space to meet Jaime, a college student working as a summer intern for a technology start-up and a more familiar figure in narratives about digital technologies and work. When it was time for our interview, I watched as Jaime removed his noise-cancelling headphones from underneath the hood of his sweatshirt. He peeled himself away from his computer and spun around in his ergonomic chair as he looked for me across the open office that, despite being crowded with people, was eerily quiet. As he approached me, he stretched his arms and legs and explained that he was a little stiff after sitting in the same position for several hours, completely absorbed in his work building a website for one of the start-ups that shared this large coworking space in Washington, DC. Jaime was soft-spoken and contemplative, but his quiet demeanor belied a forceful sense of ambition and drive for status that motivated his goals as an "entrepreneur" and much of our talk about his digital technologies. When I asked him to think about the things he liked the most about these technologies, he wrote "ability to self-promote and build my brand" first on his list. He explained:

> I want to get out there, I want to have my name known . . . and one of the ways that helps is to know people and for people to know you. . . . So this ability to promote myself on Twitter, on LinkedIn, Facebook, my personal website, it's all for me a plus. It's also cool that you generally can feel a connection so . . . I know for a fact that . . . I'm two steps away from being personally connected to the president . . . and when I look on LinkedIn, I can see my second, third, levels of connections. That helps me feel closer to these people and it also motivated me because I feel like I can climb that ladder easier, it's open space.
>
> Julia: Can you tell me more about how it makes you feel closer to them?
>
> It's not so much like a personal feeling, it's more of a "Wow, I'm actually not as far away as I thought!" kind of feeling. We have this tendency of thinking, of seeing big-shot CEOs as superhuman, sort of another level, but in a way this humanizes them for me. It brings them down to this level where they're just another person that I can connect with, and I might even know someone that can help me do that. So it brings them down a little, and elevates myself as well. It makes me feel as if I can get up to that level.

Jaime explained that the visibility he can achieve through social networks and his personal website is essential to building his reputation as an entrepreneur and to his status within the technology business elite. As he was an educated, highly skilled worker entering a fiercely competitive field, it was exciting for Jaime to be able to visualize the "open space" between his existing social and professional connections and the steps in the "ladder" to becoming a successful tech entrepreneur. Through the social capital of his "connections" and, by being highly visible in these networks himself, Jaime hoped to "elevate" himself to the level of the "big-shot CEOs" he admired and achieve his goals. By being able to see how his social networks connected with others through sites like LinkedIn, and the publicity he could achieve for his own personal "brand," Jaime relied on different kinds of visibility to navigate the fast-paced and cutthroat world of technology start-ups and achieve his professional goals.

We had decamped to a nearby coffee shop for the interview, but on our way back to his office, Jaime and I ran into his boss, the founder of the small start-up, who had come out of the elevator and stopped to chat with us in the lobby before running off to an off-site meeting. He jokingly asked me what Jaime had said about him in the interview, and Jaime volunteered that we had talked more about his social media than the company. "Oh! You mean my free marketing?" he said to me, smiling. "I always tell them, 'Whatever it takes to get it done!' They have to be able to code, but if they're Facebooking and Instagramming and whatever, then it's good for all of us, right?" On the way up the elevator, Jaime explained that he frequently posted positively on his own social media about what he was doing at work, saying that "When my product takes off, even if I'm not here anymore, I can point back at these posts and say, 'I was there, I did that!'"

Jaime valued the way his devices allowed him to engage in a form of visibility called "self-branding," or "the strategic creation of an identity to be promoted and sold to others."[4] This kind of visibility is a strategy the elite Silicon Valley technology workers that Alice Marwick studied used to gain social status in an uncertain economic climate that places a high value on individual responsibility and meritocratic achievement. Just like Jaime, these elite workers used social media and personal websites to "build their brand" and exert a measure of control over their professional lives amid the flexibility and riskiness of their jobs and companies. Although their start-up may go belly-up tomorrow, or their short stint as an intern or other temporary collaboration ends, strategies of visibility that publicize their reputations allow them to accumulate a portable form of professional social capital that helped them hedge against that uncertainty and create a longer-lasting professional identity.

As Jaime's and Charlie's stories reveal, digital technologies were essential for both high- and low-wage workers to find and secure paid work and to construct a meaningful sense of identity while coping with the risks that run through their respective labor markets. However, despite the similarities in the ways these two men use their devices, we often think about knowledge workers like Jaime as highly skilled and savvy users and don't notice, or actively discount, the equally sharp skills of so-called low-skilled workers like Charlie.[5] As digital technologies become increasingly necessary to workers across the labor market, economic conditions of insecurity and widespread perceptions of economic instability have created similarities in unexpected places as dissimilar types of workers use their digital technologies toward similar ends.

The Digital Hustle

In many ways, we might predict that workers like Jaime in the entrepreneurial technology sector would embrace the personal risks that accompany the potential for immense financial opportunities. His excitement about using visibility to craft his brand and measure his proximity to power and prestige through his social network connections might translate into his own wealth, and his strategies are bolstered by an ideology about risk-taking in Silicon Valley that glorifies individual responsibilities for success and failure.[6] However, far away from the wealth of Silicon Valley, I found that similar strategies were also essential to some of the most vulnerable workers I interviewed. For the snow shovelers, landscapers, and other low-wage gig service workers in this study, digital technologies were also essential to their ability to find, secure, and coordinate work in formal and informal labor markets. I call this ensemble of strategies a "digital hustle" to evoke the sense of constant movement, vigilance, and savvy resourcefulness that both high- and low-wage workers communicated throughout our interviews. I use this term as a way of understanding these strategies not simply as a group of work practices oriented toward the market, designed to find work, but as a form of labor in themselves as well as a set of creative and meaningful practices oriented toward the self and others. These strategies are a response to insecure job markets, and while they sometimes provide a meaningful sense of pride and identity, they also make new demands on workers to labor with and through their technologies.

The digital hustle is a complex project that individuals mount to creatively navigate insecure labor markets. This project is undertaken in the hopes of securing paid work and managing existing gigs, but the work of the digital hustle itself is unpaid. Unpaid labor done to prepare for paid work is nothing new, such as the "aesthetic labor" of high-end retail workers and bartenders.[7] Digital media companies like Facebook and Google have long relied on the free digital labor of people clicking on ads or posting about their lives. In the feminist tradition of drawing attention to unpaid work that is unrecognized but essential to economic life, digital labor scholars have pointed out the essential labor of volunteer chatroom and online discussion moderators, bloggers, and those who provide free personal content to online sites like Facebook and LinkedIn.[8] But little attention has been paid to the unpaid digital labor that scaffolds paid work. The unpaid labor of the digital hustle exacerbates inequality by requiring the most insecure among us, those juggling many gigs, to do more unpaid work than those whose income comes from a single employer.

The term "hustle" has a complicated political history. From Motown through disco and hip-hop music to its appropriation by entrepreneurs in Silicon Valley, the word's wide resonance appeals to those currently navigating flexible and insecure labor markets.[9] However, as Lester Spence points out, the meaning of the term has changed over the twentieth century: "Whereas in the late sixties and early seventies the hustler was someone who consistently sought to get over, the person who tried to do as little work as possible in order to make ends meet . . . the hustler is now someone who consistently works."[10]

Once associated with sex workers, pool sharks, and others engaged in "gray" market or criminalized work, it was formerly understood as a part of informal or "under-the-table" work in racialized labor markets.[11] Loïc Wacquant argues that the hustle "stands in structural opposition to that of wage labor in which, at least in theory, everything is legal, recognized, regular and regulated, recorded and approved by the law, as attested by employment forms and wage slips."[12] However, for the residents of the poor and Black neighborhoods in Chicago that Wacquant studied, hustling to make a living was "known and tacitly tolerated by all because [it was] both banal and necessary."[13] From this perspective, the hustle is both irregular, from the perspective of the formal labor market, and unremarkable, in the experiences of impoverished and marginalized people. In these socioeconomic contexts, formal and informal temporary, seasonal, and independent gigs are a regular, if not stable, feature of paid work.[14] While some kinds of hustles are racialized, criminalized, and stigmatized, the hustle is actually the normal result of a social logic that excludes people of color from formal labor markets.

The digital hustle isn't exclusive to informal markets. Indeed, the casualization of formal labor markets, especially but not exclusively for low-wage workers, has seen the spread of contingent and independent work arrangements into formal labor markets.[15] Hustling has its roots in the United States' history of racialized economic exclusion, pushing many workers of color "off the books" and into informal labor markets. However, practices like flexible and "zero-hour" contracts and temporary and independent contracting arrangements have blurred the lines between experiences of formal and informal work, making strategies like the digital hustle useful for both highly paid and low-wage workers.

Hustling isn't only a strategy for economic survival; it's also an ideology. As Tressie McMillan Cottom explains, hustling "is an ode to a type of capitalism that cannot secure the futures of anyone but the wealthiest."[16] The hustle celebrates a ceaseless, sleepless, restless dedication to earning money at any cost and has been at the center of rap and hip-hop music for decades. As the Great Recession in 2008–2009 wreaked havoc across many income brackets, the word began its slow creep onto coffee mugs, sweatshirts, and office artwork.[17] The beginning of the twenty-first century saw shifts in the precariousness of work for people who thought they were owed a more stable livelihood, whose educations and jobs were supposed to protect them from insecurity but didn't, and many of these workers found solace in the music written by people who have been coping with structural exclusion and discrimination in formal labor markets for centuries. A less generous interpretation is that White, relatively privileged workers who thought they were safe from precarious work appropriated the ideology of the hustle from Black culture as soon as they realized they might be in a similarly leaky, although certainly not the same, boat.

Hustling ideology has become synonymous with the tech industry and entrepreneurial start-up culture, where stories of young White men like Mark Zuckerberg and Steve Jobs are narrated as heroes' journeys from college dorms and garages to executive suites at their own multi-billion-dollar companies. However, while hustlers like former Uber CEO Travis Kalanick are celebrated and rewarded, at least for a time, for disregarding rules and regulations and single-mindedly pursuing profit, hustlers like Eric Garner, who was murdered by a police officer after allegedly selling loose cigarettes, are violently punished. As Michael Eric Dyson explains, the United States "praises white hustle but despises such agency in their darker kin."[18] From this perspective, it isn't enough to understand the practices that make up the digital hustle; it's essential to examine the reception of these practices. But racism doesn't only shape the reception of workers' digital hustles; the hustle itself is a product of the United States' long history of racist violence, exclusion, and discrimination against people of color in formal labor markets.

For both the high- and low-wage workers I interviewed, their digital technologies were essential to cobbling together patchworks of paid work, juggling multiple work schedules, and coordinating care in ways that facilitated their paid work. Like a hard-working knee or elbow joins together the movements of bones on either side, the digital hustle is a kind of "articulation work" that fits together time, place, people, and tasks in the face of unanticipated contingencies.[19] These time-consuming tasks are unheralded and often invisible within organizations and become even more so when they're undertaken outside the context of traditional employment.[20] However, the digital hustle also articulates on a deeper level than these daily tasks. For many workers, the digital hustle articulates a sense of dignity and self-worth out of the fractious flow of nontraditional work.

The digital hustle is made up of at least three kinds of labor: coordination, maintenance, and compliance. These practices are adaptations to insecurity but, like most adaptive strategies, the digital hustle challenges some inequalities and reproduces others.[21] In practice, these three types of labor are more like moments in a stream of constant activity, overlapping and inseparable from one another; however, for the purposes of this chapter, I'll describe them separately and point out the places where one kind of labor may slide into another.

Coordination

Both high- and low-wage workers used their digital technologies, especially their smartphones, to manage their clients' schedules and their own. This was no simple task and often involved aligning multiple types of schedules, including schedules for their gigs, for hourly or salaried work, and for the rhythms of family and social life.

Like many of the workers interviewed for this book, Lisa earned money by cultivating a diverse set of gigs, which, while providing some stability in income, multiplies the time she spends scheduling and coordinating clients and their demands. Lisa is a college grad living in the Bronx—while she was looking for a "more stable" job in writing or editing, she was hitting the digital pavement to look for child and elder care, home organization, and other low-skilled errand work on TaskRabbit, Care.com, and Craigslist. At times during our interview, it was difficult for me to keep track of all of her different streams of income.

At the time of writing, TaskRabbit requires "taskers" to log in to indicate blocks of time when they're available (8 a.m.–12 p.m., 12 p.m.–4 p.m., and

4 p.m.–8 p.m.), which they cannot alter. Taskers can choose whether or not to be hired for "same-day" gigs and whether they want to work "off hours," meaning after 8 p.m. This presents a challenge to Lisa as she will often be coordinating her work running errands sourced through TaskRabbit with her work as a professional organizer and babysitter. At times, she may be free all morning but have an appointment with a long-time client at 12 p.m., which means she has to mark herself as unavailable for the entire first block. She told me that, at first, she tried to take a risk and mark herself as available just in case she may get booked for a morning gig that would wrap up in time for her afternoon appointment. However, she quickly found out that if she should get booked at 11:30 a.m. and have to turn down the gig, the metrics displayed on her TaskRabbit profile—which are linked to pay rates and how highly she appears in ranked searches and also assure potential clients that she's trustworthy and will do a good job—suffered and it was difficult to recover her reputation so she could continue getting booked. Now, she uses the blocks of time TaskRabbit requires her to use to schedule the rest of her gigs and tends to think about her work days within the four-hour increments set by the platform.

Lisa credited her ability to manage her complex schedule to her love of organizing: "I think it'd be very difficult if you weren't OCD [obsessive-compulsive disorder] like I am, I just love organizing." She explained her elaborate system for creating her schedule:

I have alerts from TaskRabbit popping up all the time and I check them right away, as soon as I can, but I only check my personal email twice a day, so in the morning and then whenever I'm done with my day. And then when it comes to the gigs that come in through my email, I try to do things very quickly. So if I get a lead, I put them in my contacts immediately. . . . I probably have like a few thousand [contacts]. And then I use my Google Calendar for the things I'm actually going to do. So, when I'm on the phone confirming with someone, I'll put it straight into my calendar while they're talking. Then, with the babysitting stuff, that's more regular with two families so they just text me and right when I get that text it goes into my calendar first, then I write them back to say "yes," otherwise I'll forget to put it in the calendar. I also confirm with all clients the day before, because if I don't confirm, and then the person doesn't let you in or people cancel, I don't have time for that with all the moving parts I have going on.

Lisa coordinates a nearly constant stream of appointments received through an app, emails, phone calls, and text messages with her calendar app to transform an ever-shifting jumble of one-off and regular gigs into an orderly schedule of paid work. Lisa used the clinical term "OCD" colloquially to describe her over-the-top organizational style, but, as Harry Braverman points out, "that which is neurotic in the individual is, in capitalism, normal and socially desirable for the functioning of society."[22] The complex coordination labor Lisa has to perform to keep her gigs running smoothly is a requirement, not a personality flaw, to earn a living in contingent work.

For high-wage workers, coordination labor more often involved multiple clients rather than multiple streams of income. Travis was a self-employed communications consultant living just outside San Francisco. He had a gentle and soft-spoken manner, and, after spending thirty years working in education nonprofits, which "ate up my time and spit me out the other side," he decided to try striking out on his own before retiring. For Travis, coordination labor often involved learning the preferences of his clients and anticipating their quixotic desires. He explained:

> I don't have control over my schedule, that's a huge misconception people have with freelancing.... I always yield if a client wants to meet at 2 p.m. on Thursday, but I had scheduled that block in my calendar to get some work done, it doesn't matter, I'm there! I schedule my life around my clients, basically, but I've gotten better at . . . kind of passive aggressively *suggesting* times that I know will let me fit everyone in.

Travis recalled a situation in his first year of freelancing when he was working with a very demanding client and devoting so much time to them that he lost another client because he wasn't able to meet when they needed him. "That's when I realized I had to be more active in coordinating people, like I couldn't just let one client walk all over me." He developed a routine that involved strict rules about immediately entering meetings into his digital calendar, which synced across his phone and laptop, as well as never sharing, or allowing digital access to, his calendars with his clients. Even though entering in meetings individually was time-consuming, Travis admitted to being a "control freak" and worrying that someone might "mess up the system" by trying to alter or add meetings. Like Lisa, Travis described his tight grip on his schedule pejoratively, attributing it to a fluke of his personality rather than a necessity of his work.

For some, schedules of paid work were only one piece of an even more complicated dance of children's activities and the work schedules of spouses and others involved in the care of children, which also required deft organization facilitated by digital devices. Independent and contingent workers often operate outside standard 9–5 "working time," which can be helpful for coordinating childcare and accommodating mid-day appointments and errands but also raises the level of complexity and unpredictability of household scheduling. As I'll explore in more depth in Chapter 4, this kind of labor is gendered, often falling to women.

The labor of coordination lives in the software and practices workers use to create a smooth flow of work from the staccato rhythm of gigs, but it also lives in the work of carving out or holding space for the unexpected. Coordination labor, perhaps more than the other parts of the digital hustle, has been intensified under postindustrial forms of "flexible capitalism." As Benjamin Snyder explains, while twentieth-century industrial capitalism conceived of working time within the rhythms of regular, uniform, and predictable 9–5 office and shift work, postindustrial flexible capitalism is ruled by the "just-in-time-ness" of quixotic consumer desires, fluid and rapidly changing financial markets, and complex global flows of goods and services.[23] This shift entails a different conception of working time that's characterized more by the jazz musician's sense of perfect timing and adaptability than the punch clock's mechanical countdown to quitting time.

Lisa contrasted her experiences working many gigs with her friends who work more traditional schedules:

> I talk to my friends who are nurses and teachers and it's just so different, their life is so structured, they have normy jobs . . . you know, like more normal. They know how much they're going to make and their salary increase, and it's nice, but it's so different from my experience . . . it would be nice to know I had a nice check coming in every month, but at the same time it's kind of satisfying knowing I can make all these gigs work for me, like when I have a full day of gigs and someone cancels and I figure out how to make it all work, or when it all goes smooth, like someone goes from emailing me from Craigslist to a text, to a phone call, to a gig, to a Venmo payment it makes me feel good, you know? It's weird but, I get a boost from that, like I'm still beat at the end of the day, but it feels good.

After college, Lisa transitioned from holding more "normy" jobs like her friends to doing freelance work because she wanted more time to write and pursue creative work. As she described earlier, the insecurity of her paycheck is frustrating, but when coordination labor goes right, turning a stream of complex scheduling into income, it makes her feel good. When coordination labor goes right, it can be energizing and exciting, but something as simple as a cell phone left in a car instead of a pocket shows just how tenuous this labor is and how contingent it is on the affordances of digital technologies.

Maintenance and Cultivation

While the work of the digital hustle often takes on a frenetic, energetic pace and is narrated in stories of clients heroically won or projects delivered just in the nick of time, there's another register that operates in the background of these more eventful moments: the two-steps-forward-and-another-one-back work of maintenance and cultivation. Once the personal websites have been made or the Craigslist ads have been posted, maintenance labor is the slow work that sustains online and offline reputation and keeps technology working smoothly, while cultivation labor is the more forward-looking check-ins with past clients and attendance at networking events. Charlie's description of reposting his Craigslist ads before our interview so that "people continuously see me" is an act of maintenance labor in his digital hustle. Similarly, his reminders to each of his clients to rate him through Uship.com are an example of the gradual and cumulative logic of this type of labor.

Maintenance labor isn't only necessary for reputations created from five-star ratings systems but also to sustain one's reputation as it moves across online and offline contexts. Constantine, who ran his own nonprofit for Black youth in Washington, DC, and worked as a diversity and nonprofit consultant, told me that it was time-consuming to attend networking events that had uncertain payoffs for his work:

Do I go to the hundredth Black Entrepreneurs or nonprofit meetup I get invited to? Or do I put that three or four hours into the work that needs to get done right now? These events are all over social media now so it's not just who you meet at the event but the publicity that comes after, too. You don't want to be the one who missed that great event that people are tweeting pictures, inspirational quotes from the speakers, or whatever big-shot donor or investor they got to meet, but if nobody shows up, I'm sitting there like, "I could be solving five problems instead of sitting here." You don't want the scene to forget you, but you have to do the work, too!

The trade-off between doing the maintenance labor or being visible through photos and tweets to make sure his networks don't "forget" him and spending time "solving problems" for his organization or clients is a delicate balance for Constantine. While many scholars have examined the phenomenon of self-branding, few have observed the work that comprises the background of this slow and cumulative process.[24] This work—less brazen self-promotion and more of a "responsibility for being pleasing, reaching out, connecting with others"—is what Nancy Baym calls "relational labor."[25] The creation of a personal brand or online reputation depends, in large part, on the persistent, mundane, and slow accumulation of ratings, reviews, and other traces of interactions that are eventually made visible through websites and labor platforms. These nearly invisible and intensely personal practices may feel insignificant on their own, but when gathered together, these "discounted labor practices" are an important type of labor in contemporary modes of connected work.[26]

These practices of maintenance labor are premised on access to the Internet and working digital technologies; however, they also secure these background conditions. Working in a rural part of New York State prone to ice storms and other bad winter weather, Mary, a virtual executive assistant, pointed out:

> There have been times when my Internet has gone out and I'm completely helpless. I can't produce the reports and have access to the files that my bosses need. When they're headed into a meeting with the president of the company, that's a problem! Sometimes it's bad weather and . . . what're you going to do? I've had to take my computer and literally go work at Dunkin' Donuts, or I'll go to my neighbor's house or something.

Mary's ability to do her job depends on broadband Internet access in her home, and, when it's not available there, she has to work to find access elsewhere. As Amy Gonzales points out, "access to digital technology is not a permanent or categorical state but rather an ongoing experience of labor, negotiation, and coping."[27] The work of maintaining an Internet connection doesn't always require such quick thinking but also persists for Mary in the monthly ritual of paying her bills for home Internet and cable, occasionally haggling for better rates or shopping around for more reliable services. This kind of technical maintenance labor depends on a relatively sophisticated level of technological knowledge and also on the often hard-earned understanding of how sites like Craigslist and Uship.com work. As Charlie described:

> It took me a while to learn how often you can post ads on Craigslist before you get flagged. I don't know the exact formula or whatever, but I was really pushy about it when I first started and would post every hour or so, and I got a bunch of posts taken down because of flags. And then, someone told me about Uship and I wasn't sure how to use it at first, but eventually after a couple of times, I figured out you really have to have some kind of ratings on there to get the higher paying jobs. If you don't have some reviews on there, the bigger fish aren't going to take a chance on you. I don't even think everyone can see your profile on there until you have reviews.

Charlie developed theories about the inner workings of Craigslist and other labor platforms through trial and error, slowly making enough mistakes that he was able to understand the logic of success in these socio-technical markets. The maintenance labor of slowly cultivating an online reputation depends on these provisional understandings about the proprietary technologies that facilitate Charlie's relationships with clients.

Compliance

If maintenance labor follows a slow and accumulative logic, compliance labor often feels wrenching and effortful. Compliance labor isn't aimed at learning the rules of the game but at the everyday work of constructing a self that reflects those rules. As Salena, a communications consultant at an education nonprofit, described her job, she slickly pitched the value she added to the organization as a "brand strategist" who was an expert in "social media engagement." Offhandedly, she said, "Except, I don't apply it to myself at all. . . . I really don't care enough to do that, it's too much work for me, and I really don't care that much." As Salena would go onto describe, the work of creating a personal brand goes far beyond creating a website to advertise her services:

I'd have to comb through all of my tweets and Instagram pictures and not only make sure they're appropriate, like for work, which of course they already are, but also that they're "on-brand," which means making sure they build a kind of organic or seemingly natural story about who I am and that I'm supposed to be an expert in whatever thing. It's really time-consuming, going through everything and thinking, is this something that adds to my brand? Or is this just a random cool thing I want to post because I read about it? And how much random stuff should I post? If it's all "on-brand" then it's too curated, too one-track mind and that's boring, it doesn't look natural.

Ironically, the creation of a "natural" personal brand is, in fact, in compliance with a set of style rules Selena has internalized and is also, although engineered to look effortless, quite effortful. Marwick calls this kind of labor "brand monitoring," or self-surveillance that internalizes the perspective of one's audience to continuously edit and construct a "safe for work self."[28] Compliance labor however, goes beyond the monitoring of one's social media content and includes the work it takes to follow the rules of digital engagement as set out by clients or the labor market, whether explicitly articulated, assumed, or imagined.

Unlike Salena's imagined audience for her "authentic" personal brand, Lakisha's compliance labor was done for an audience of one, her shift supervisor. For her seasonal job serving appetizers and drinks in the hospitality boxes of a large football stadium, Lakisha was explicitly forbidden to have her phone anywhere on her body and was required to keep her phone in a locker located in an employee locker room in the basement of the stadium. This rule was spelled out in the manual she received in an email (which she was expected to read before her first day) and at orientation on her first day, and it was repeated by her shift supervisor at pregame meetings each day she worked. When I asked her why this particular rule was so important, she explained that there were often celebrities in the boxes, and her employer wanted to keep the workers from disturbing their privacy by taking photos of them. Lakisha was less concerned with the celebrity sightings and more concerned with being able to receive updates about her infant son, usually in the care of a rotating cast of aunts and cousins, whoever had some free time to look after him when a shift opened up. She worked in the uppermost levels of the stadium and, if she jogged, it took nearly ten minutes to wind her way down into the employee locker room in the basement to check her phone, making mid-shift check-ins difficult, though not impossible. While these communication blackouts are something mothers dealt with long before cell phones, their widespread use has engendered expectations of constant availability, especially from mothers, even if some workplaces don't allow the fulfillment of these expectations.[29]

One night, when her son was teething, Lakisha was getting a steady stream of texts with questions and concerns from a teenage cousin who couldn't remember if the baby's teething ring should be hot or cold, what dosage of Tylenol she should give, and whether all that drooling was normal. Lakisha knew everything would be fine but didn't want her cousin anxiously making a mistake; she ended up relying on her coworkers to cover for her so she could dash down to her locker multiple times to check in and answer questions. "I dropped five pounds that night!" she recounted with a laugh, "but I never tried to sneak my phone, can't afford to make a mistake at this job." But, inevitably, there are times when she can't make the dash through the stadium to check in and has to "just take deep breaths and tell myself, 'The baby's gonna be fine, they'll figure it out' and keep myself busy so I'm not worrying about it." Lekisha's compliance labor was physically taxing but also required work on her emotions. If coordination and maintenance labor are carried out on schedules, websites, and ratings systems, compliance labor is disciplinary— not in the sense that it's violent or punishing but in the sense that it's carried out on the self and rewards those who can conform their subjectivities to the demands of precarious work.[30] This type of labor highlights how conformity may appear to be the path of least resistance, but compliance can require a great deal of effort.

Inequalities run through these three types of labor. The power that workers have to resist compliance or to adjust client demands to fit a childcare schedule has everything to do with their social status and power. Similarly, both high- and low-wage workers had maintenance issues at some point, but class ultimately shaped the parameters within which workers dealt with the inevitable breakdown of their digital hustles and marshal resources to their aid. These issues are central to the following two chapters.

Between the Hustle and the Heart: Crafting an Identity

The digital hustle isn't solely oriented toward the marketplace. Economic motivations were central but don't fully explain Charlie's pride, Jaime's excitement, or Mary's frustration. In his study of the economic life of a Chicago public housing project, Venkatesh observed that residents understood "the hustle" as both a set of survival strategies and a means of crafting an identity. For residents, hustling was "simultaneously about adapting to material constraints and attempting to reproduce a self-efficacious, meaningful existence."[31] In this sense, what's notable about the hustle isn't only *how* it's done but also *what it does* for the people doing it. The digital hustle is foundational to gig workers' ability to create and sustain a sense of dignity and self-worth across different streams of income and clients and without traditional employers. Instead of finding self-worth in raises, promotions, or good evaluations from mangers and bosses, they're finding dignity in successful digital hustles.

Tori and I met at a Verizon store in a suburban strip mall in Rancho Rio, California. I noticed her Brooklyn accent as she joked and gently chided her friend about the privacy settings on her social media accounts while they waited for a screen repair on her phone. Over pizza later that day, she and I talked about what she was proud of about her career as a babysitter, nanny, and household manager:[32]

> People [her clients] who already know me . . . they tell all their friends, and referrals typically come through text message. So, I'm always on my phone, checking messages. It's my business, it's how I survive. But, I'm my own person, I've never had to reach out to any of those [agencies]. They're always like "Well, we pay $13 an hour." [Scoffs] Even when I was just starting I made $15 an hour. I know what I'm worth, I know my rate, and if my clients want to pay it, they pay it. I get better rates than the agencies offer, and I do it all on my own. I run my own business and I don't need them, but if I fail it's on me. It makes me feel good, you know? Like, I pay my own health insurance because I'm good at this, not just with kids, but with the clients, these people hand me their credit cards, their keys, and their kids. . . . I think it's because I'm good with the referrals, I always text right back, I'm always prompt and professional, you know? And they trust me, and I make money sure, but I'm also helping them, helping their families, right?

Tori's digital hustle, including her "prompt and professional" practice of texting clients and referrals back quickly and building trust with her clients, is necessary to her "survival" in the carework industry. However, the labor of her hustle is not only necessary, it's also good. Tori's pride at her independence from nanny agencies—who often act as matchmakers between careworkers and families—her ability to ask for the rate she feels she's "worth," and her ability to afford health insurance are all a result of her hustle and constitutive of her dignity as a professional in this field.

The digital hustle is a craft practice of the gig economy. Often associated with skilled and masculine trades like carpentry, masonry, or butchery, the idea of craft is not often applied to digitally mediated work.[33] Digital technologies are often associated with the exact opposite of craft, namely deskilling, automation, and the loss of autonomy over manual work. However, as Randy Hodson points out, "a defining characteristic of craft workers is intense pride in their work. Long experience with substantial technical training produce[s] a level of expertise that workers experience as a source of great pride. The personal identities of craft workers thus tend to be firmly grounded in their occupational skills and activities."[34] While the technical training of craftworkers is usually assumed to include formal and organized courses or apprenticeships in manual trades, the long process of learning the ins and outs of different platforms and websites, or the habits of mind and hand required to vigilantly and professionally attend to texts and emails from clients and bosses, constitute a significant investment in technical training by practitioners of the digital hustle, although they're seldom acknowledged as such, and a source of "intense pride" and personal identity.

Tori is acutely aware of the downsides of her hard-won independence, noting that she's also the only one to blame for her failures. This insight isn't a theoretical one for Tori. She told me about an experience where she accepted a highly paid job from a notoriously difficult client, who interviewed Tori to be her nanny after three others quit after less than a year working for her. Tori had spoken to her former nannies, who told her that this client was distrustful, using cameras in every room in her house to surveil their every move, often asking about their decisions in specific situations throughout the day when she got home in the evening. But, never being one to shy away from a challenge, Tori accepted the job, figuring that the challenge of creating a trusting relationship would be worth the eventual payoff in lucrative referrals and networking in her clients' social circle of highly paid professional moms. Tori was confident in her ability to work with the unique demands of what she called "high-demand, high-profile" clients, reasoning that her communication skills and experience would allow her to overcome the obstacles the other nannies had faced to establish trust with this difficult client. However, after just three months of late nights, constant personal surveillance, and a steady stream of text messages and phone calls throughout the day questioning her judgment, she put in her notice.

After this calculated risk didn't work out, Tori could have felt despondent and defeated or anxious about having lost her main steady source of income. But when I asked her how she thought about the aftermath of her decision to quit, she explained:

> I mean, it happens ... but, there's really never any time or day where I can't get work. I might go ahead and text Mary and say [moving thumbs across an imaginary phone keyboard], "Hey, Mary, I'm open tonight. You need any help?" And Mary will be like "Come on over," and I'll be like "Making a U turn, on my way." If I see that for some reason I don't have any clients on that particular day. . . . I will reach out to particular clients that I like and say I'm available and nine times out of ten it works. Clients usually say they can't use me same day but are like "Can you come Thursday?" And then I'm like "Hell, yeah." I'll do that for two or three clients for every one that cancels and then I end up being booked for two days instead of one. That's how the money gets in my bank. . . . I mean, did it suck to lose the money from that gig? Yes. But when I can just jump on text and get people to book me right away, it's like, I'm clearly better off without it.

Tori's confident and pragmatic assessment of her earning power paints a picture of her ability to navigate a jungle gym of economic opportunities; when she briefly loses her grip on one, she reaches out with her digital hustle to find two or three other handholds to grab onto as she zigzags across her career. Tori's ability to instantly send texts to tell her clients she's available allows her to nimbly change her plans to take advantage of other opportunities. She's "better off" without the security and predictability of a bad job when she's able to quickly use her phone to create her own opportunities elsewhere. Autonomy is widely understood as a cornerstone of dignity for highly skilled professions.[35] The romance of risky work, the excitement of uncertainty, the creative problem solving required to keep things afloat in rough waters, the autonomy to try new things and potentially fail—these are qualities of work that many scholars point out attract high-wage knowledge workers to work for start-ups or other uncertain ventures.[36] But these same pleasures are not often recognized in low-wage work. Instead, the long-held assumption is that while highly skilled and educated workers are allowed the privilege of opting for insecurity (albeit perhaps against their own financial best interests), lower wage workers don't choose but are forced into similar conditions of precarity and insecurity and would prefer stable and predictable work. While these assumptions may hold true for some, presuming their universality has occluded a deeper understanding of the reasons workers choose independent and contingent work. Tori's case illustrates that low-wage workers are not passive victims of labor market polarization but actively shape their experiences of work and create security in environments that are often hostile to their autonomy.

For Tori, her digital hustle helped to produce a sense of autonomy in a tough situation. Her ability to get work, earn the trust of her regular clients, and rebound from failure wasn't merely the instrumental payoff of her savvy use of her digital technologies but a performance of security in an economic environment where her next job was uncertain, her earnings were somewhat unpredictable, and the measure of her success was elusive. Tori can't rely on regular reviews of her work from a manager or track the development of her skills in terms of regular pay increases or promotions. Independent and contingent workers don't have access to the "badges of ability" that characterized previous eras of labor and use their digital hustles as indicators of their worth and dignity. Richard Sennett and Jonathan Cobb write that societies award "badges of ability" that recognize the qualities and abilities that make someone worthy of dignity and respect and mark individuals apart from the crowd and "bestow the right to stand out as an individual."[37] For Tori, and for many of the interviewees in this book, who were stitching their livelihoods together outside the bounds of single organizations and employers, the successful execution of the varied labors of the digital hustle—maintaining a full schedule of constant activity, savvy management of schedules and opportunities, effectively leveraging social networks to one's advantage—were the markers that broadcast to themselves and others that they weren't just surviving but instead thriving in insecure circumstances. In these ways, workers like Tori don't just rely on their digital hustles to find work; their hustles also work on them.

Work is one of several major social institutions that shape collective symbolic boundaries between good and bad and worthwhile and worthless that order our social worlds.[38] However, when workers' experiences aren't defined by a single employer or company, then the work practices and sense of productivity that follow them become important sites of identity and judgments of moral worth.[39] Tori's, Charlie's, and Jaime's digital hustles came to mean more than a means to an economic end. Charlie's pride in his high rating,

Jaime's thrill at being one step away from the president through social media, and Tori's confidence in her ability to rebound from risks are signals that the work of the digital hustle is foundational to their identities as workers between the boundaries of single employers and organizations.[40]

Conclusion: United by Hustle

The workers I interviewed clearly relied on their digital hustles to negotiate a sense of worth from the flow of multiple jobs and gigs and creatively empower themselves from within difficult labor market conditions. But what kind of a solution is it? The digital hustle is a private and individual solution to dilemmas that were created by public and collective choices over the past several decades. These strategies are a response to the polarization of the American labor market, the shifting of risks for career advancement and success onto the backs of individual workers, and the disinvestment in public safety nets. From this perspective, the digital hustle could be seen as a "technology of the self," not because it makes use of digital technologies but because, as Foucault theorized them, these activities are "micropolitical rituals and practices that configure particular forms of subjectivity . . . [and] serve important political and ideological functions, chiefly by aligning forms of subjectivity with the needs of the neoliberal economy."[41] Through the daily, routine activities of the digital hustle, workers' sense of self is disciplined to conform to what's best for the market, not necessarily for their own ends. So, while the digital hustle is central to how independent and contingent workers define themselves, it also produces a self that keeps workers trapped in the logic of entrepreneurial capitalism.

Aspects of the digital hustle are disciplinary and condition workers who share very little else in terms of their social class or occupation. However, this perspective might lead to the incorrect assumption that the labor of the digital hustle conditions all workers who use it in the same ways, evoking similar reactions, emotions, and conflicts. In reality, these strategies construct the field of contingent work in vastly different ways for high- and low-wage workers. While I found that both high- and low-wage workers used their digital hustles to find work and advance their careers, as I show in Chapters 2 and 3, these practices ultimately reproduce inequalities embedded into conditions of their work.

In highlighting the similarities in the labors of contingent and independent workers across classes, the digital hustle illustrates that the personal address of entrepreneurial capitalism isn't confined to the elite and demonstrates a "community of practice" where it may not be readily apparent and even obscured by current labor politics.[42] Bringing together the labors of two very different groups of people shows that the labors of the digital hustle unite a divided labor market. In the midst of a widening gulf between high- and low-wage workers, shared experiences of the digital hustle suggest hidden similarities between independent and contingent workers. In this unique moment in US labor history, social conditions beyond the traditional triumvirate of inequality—race, class, and gender—are fostering similarities between otherwise dissimilar groups. In this case, the destabilization of paid work has created conditions wherein digital technologies enable high- and low-status workers to piece together patchworks of paid work and help them construct an identity as worthy workers in insecure circumstances.

However, while these work practices confer a degree of legitimacy and reassurance to contingent workers across the labor market, high- and low-wage workers hustle under very different circumstances. The "badges of ability" of the digital hustle call into question the kind of legitimacy these strategies confer. How durable is the reassurance? What differences lie behind the similarities in these workers' hustles? What do different kinds of workers have to do to secure these badges? The next two chapters will examine these questions from inside the experiences of low- and high-wage workers, respectively.

2

After Access

The Costs of Inclusion for Low-Wage Workers

On the day we were scheduled to meet, Mike was fifteen minutes late for our interview and not answering my texts or phone calls. I was just about ready to leave when he burst through the door of the Dunkin' Donuts in Washington, DC, where we had agreed to meet, full of apologies. He sat down, and, as he put his phone on the table, it came to life with sound and vibration. Mike said that after spending over $100—the equivalent of ten hours' pay—on a new phone that had ended up in a snowbank after a friendly tussle with a friend, he had to scrape together more money to purchase yet another phone. Because of the unanticipated cost, however, he wasn't able to keep up with payments to his mobile phone carrier.

Mike had two part-time jobs, one in a medical supplies warehouse and another as a day laborer for a home repairs business, both of which required him to commute for over an hour from where he lived in Maryland. For both jobs, his bosses would call to check his availability, often only a few hours in advance. With his lack of consistent phone service, this posed a logistical challenge and had led to his habit of spending days off hanging out at local restaurants with free Wi-Fi. He used MagicApp, a free app downloaded onto his phone, which allowed him to make and receive phone calls using the free wireless Internet he can access at fast-food joints around the city. He told me how, on a day he hadn't yet been scheduled to work, he'd head to the McDonald's down the street from his house to call his bosses and check Craigslist to find other gigs.[1] For others, this might be a problem, as McDonald's and other fast-food chains that offer free Wi-Fi have started enforcing rules around loitering designed to keep space open for paying customers and to keep people like Mike from free-riding on the free Internet (see Fig. 2.1). However, Mike knows some of the employees and they let him hang out and use the Internet without hassling him to buy fries and drinks every thirty minutes, allowing him to stay past the time limit.

Left to Our Own Devices. Julia Ticona, Oxford University Press. © Oxford University Press 2022.
DOI: 10.1093/oso/9780190601288.003.0003

Fig. 2.1. Photo of sign posted in McDonald's (Washington, DC). Author's photo, 2016.

On the day of our interview, his train was running late, and although the train stations are supposed to have Wi-Fi, today, like many other days, the Wi-Fi wasn't working, and Mike couldn't text to tell me there was a delay. Mike takes the same risk every day when he coordinates his work schedule; spending a few hours at the McDonald's or the library coordinating hours at his two gigs and piecing together other paid work by picking up gigs on Craigslist and through friends on Facebook, and then heading off into the city for the day. Mike's digital hustle, a set of strategies described in the last chapter as common to both high- and low-wage gig workers, required that he have consistent connectivity, which he didn't, creating a host of challenges that required creative solutions.

Among low-skilled men of color like Mike, strategies of "foraging" for work from multiple sources is increasingly common, but Mike also relied on foraged connections to the Internet throughout his daily commutes around the city—a train station, a city park hotspot, a Starbucks next to the upscale condo he was helping remodel, and the Wi-Fi password of a friends' neighbor.[2] While Mike was looking for oases of connectivity, he was surrounded on all sides by people apparently awash in ubiquitous Internet access—the teenagers on the bus next to him scrolled through social media and streamed TV shows without a second thought, his bosses called from the hardware store with last-minute requests or inquiries about his arrival, and friends texted adjustments to their evening plans.

Mike has developed tricks for padding his schedule to account for DC's unreliable transportation system; he also knows all of the fast-food joints with free Wi-Fi outside the train stations on his regular routes. If he needs to, he can jump out of the station and make a quick phone call to adjust his schedule before jumping back on the train. He has to do this because for low-skilled, low-wage manual workers like him who have short-term, informal relationships with employers, being fifteen minutes late to work might mean you don't get to work that day and could mean you never get called back to work another job. Mike said he felt he was "always runnin' behind" the messages and updates, but his race to keep up illuminates an often invisible assumption of universal connectivity among employers, coworkers, and friends. Mike's digital hustle depended on consistent connection, which he went to great lengths to secure.

While the role of digital technologies in the rise of insecure or precarious work in the United States has attracted much scholarly attention in the past several years, scholars of non-Western economies have long been writing about the "taken-for-grantedness" of digital technologies for low-wage workers in these types of work arrangements.[3] In China, low-wage gig or odd jobs workers rely on digital hustles just like Mike's to find and secure work.[4] In Ghana, young people who frequent Internet cafes full of secondhand computers digitally hustle to scam Westerners out of their money.[5] These strategies, once studied exclusively by scholars interested in emerging economies outside the United States, are now common for people navigating the constraints of poverty and the necessity of connectivity in the United States. This chapter illustrates how connectivity to phone and Internet services is essential to low-wage contingent workers' ability to find and coordinate paid work

Some might say that the widespread expectation that low-wage workers like Mike will have ubiquitous connectivity illustrates the successful closure of the "digital divide," but this frame hides the exploitative terms on which these workers have been included in the digital economy. In the previous chapter, I described how the digital divide framework not only shaped how we see the problem of digital inequality but also the solution. Access to digital technologies and the skills to use them were increasingly framed as essential ingredients of economic empowerment, presuming that low-wage workers whose jobs were under threat of moving overseas could "reskill" to better compete in a globalizing economy. [6] This framing, in turn, has focused the conversation on what low-income people "lack," or the hurdles that stand between them and the realization of their promise in the digital economy.

Mike's experiences of getting through his day with the expectations on the part of his bosses, clients, and friends that he be consistently connected makes it obvious that the problem is *not* that low-wage workers are being excluded from the promise of digital technologies. The driving force of inequality for low-wage gig workers is the terms on which they're included in the digital economy, not their exclusion from it. Expectations of constant connectivity make low-wage gig workers a target for exploitative inclusion in the digital economy.

Drawing on ethnographic work in electronic stores geared toward low-wage consumers and interviews with low-wage gig workers, this chapter will explain how the terms on which low-wage workers maintain their connectivity reinforce and amplify disadvantages they already face. First, the managers, bosses, and clients of these workers increasingly expect and rely on their phone and Internet access but don't support it in the workplace. In turn, cell phone carriers, digital advertisers, and data brokers profit from this connectivity mandate. These market actors make connectivity appear affordable but really offer poorer quality phones and more expensive plans or profit by harvesting and selling workers' data when they use "free" Wi-Fi and other online services. As this access is unsubsidized by employers and inadequately supported by federal subsidy programs, maintaining connectivity in this market levies a new tax on stagnating incomes.

Second, although low-priced plans appear to allow low-wage workers to maintain access, in fact, this access is supported by workers' "invisible labor," performed to compensate for routine disruptions to their connectivity. This work to maintain connection during disruption is invisible because it's "often overlooked, ignored or devalued" by employers and is unsupported by the many institutions that depend on it, such as workplaces, schools, and healthcare systems. [7] This labor involves a wide variety of interpersonal and technical skills and significant investments of time to work around cheap and broken phones, turned-off service, and a lack of Internet access. While some workers are entreated to take "breaks" from these technologies to preserve their mental health and productivity, for low-wage workers, disconnection isn't a source of solace and relief from their demanding jobs but a persistent condition that they must overcome to access work in the first place.

While being unable to afford to extend their monthly plans sent many workers on a hunt for alternative ways to get connected, a lack of income was not the only factor that kept them disconnected. I also explain how, in their quests to remain connected, low-wage workers, particularly Black men, found themselves harassed or prevented from using the Internet in public or semipublic spaces through racism and the threat of police violence. I argue that, to see the ways digital technologies amplify inequalities in gig labor markets, it's important to look beyond exclusion to all the ways access is conditioned by the political economy of connectivity for low-wage workers and the invisible work that goes into maintaining those connections.

Holding Open the "Digital Divide"

I mostly met the low-wage workers who would become the interviewees for this study by hanging out in the phone and electronics stores in their neighborhoods that catered to low-income consumers. Spending time in these stores, in both urban and rural areas, revealed the exploitative terms on which low-wage workers are included in the digital economy. For many low-income Americans, this inclusion begins with buying a smartphone. However, at every step of the way—from buying a phone, picking a plan, and using it to access the Internet—data and telecommunications companies exploit the mandate low-wage workers face to maintain their connectivity.

In the store of an "authorized dealer" of a national low-cost wireless brand where I recruited interviewees in Washington, DC, the management had hired a drummer and a free-style rapper to perform as pedestrians rushed by the store on their way home. The music was loud and audible inside the small store, where plastic replicas of popular smartphones, cell phones, and a couple of smart watches were glued to stands anchored to a counter around the perimeter of the store.[8] There was one replica of an iPhone; the rest were Samsung, ZTE, and HTC phones. The store's two employees were busy attending to the groups of customers clustered around different devices, often loping back and forth between customers and the cash register or a door leading to the back office and supply closet. People wandered in, often browsing around for a few moments before exiting back into the stream of pedestrians rushing past the performers on the sidewalk.

I introduced myself to Damon as he waited to find out the price of a new phone from one of the store's busy employees. Damon worked as a "floater," doing many different jobs from laundry to cleaning to maintenance at a hotel in downtown Washington, DC, as well as working at an upscale burger bar. He explained how his main phone, a budget smartphone he inherited from a friend last year, was "constantly cutting on and off, on and off." This was a problem because his unreliable smartphones were how he coordinated his work schedules for his various gigs:

> My phone kept messing up. Other than it just cutting off . . . like, it'll just turn off in the middle of me doing something and then turn back on . . . and if I charge it up 100 percent, within an hour it's already dead. . . . I'm waiting for my next check so I can get another phone. I get paid two Fridays from now.
>
> Julia: How long have you had that phone?
>
> Almost a year.
>
> Julia: Not even a full year?
>
> Nope. Every year I have to get a new phone because I just have bad luck with phones. They get a virus or something, I guess. They just start cutting out.

Damon's wireless plans, like many others, required him to pay for his phone up front. The technical problems that Damon attributed to his personal bad luck are more likely caused by the lower quality of the cheap phones he can afford. Low-end phones, from brands like ZTE, make smartphone ownership appear affordable but are prone to viruses and other hardware issues, putting low-wage workers like Damon into a constant cycle of juggling phones in various states of disrepair until they can purchase new phones, which, for many interviewees, was an annual investment of several hundred dollars.

Workers like Damon are a part of a large contingent of low-income people in the United States who depend on their phones for Internet access. In 2015, the US Census estimated that over seven million households in the United States relied on mobile devices for Internet access, one-third of whom were low-income or receive government assistance.[9] Further, while one in five American adults rely on their smartphone as their sole Internet access, for low-income adults, the proportion rises to one in four.[10] While poor people are less likely overall to own smartphones than wealthier people, when they do, they rely on them for Internet at much higher rates.[11] In particular, low-income users rely heavily on their phones for job hunting. Users from households that make less than $30,000 a year are twice as likely to rely on their phone to look for work as those from households that make over $75,000 a year.

Prepaid phones were also the most sought-after item at the remaining small shops in a downtown shopping area made up of mostly vacant storefronts in rural Mainville, New York. Wealthier residents drove forty-five minutes to the nearest mall to buy high-end devices directly from a retail store, purchasing their phones from a national wireless service provider that offered mostly annual contracts. Everyone else frequented one of several electronics and phone retailers in town. These included a neighborhood convenience store (see Fig. 2.2), a standalone electronics store that also offered bill pay services for a fee (see Fig. 2.3), and Walmart.[12]

At a convenience store in Mainville, cell phones hung, encased in hard plastic blister packs, on a spinning rack near the cash register. Brightly colored cardboard advertising that explained customers could "Save over $200 on airtime" and highlighting features of the phones hung close by. Behind the register, alongside the cartons of cigarettes and cans of chewing tobacco,

Fig. 2.2. Convenience store, Mainville, NY. Author's photo, June 2015.

Fig. 2.3. Standalone cell phone store, Mainville, NY. Author's photo, June 2015.

hung small credit-card-sized "airtime" cards of various amounts, from $29.99 through $80. Even though the cards are only "charged" with airtime upon purchase, one of the store's employees told me they had to move the cards behind the register because they were one of the most frequently stolen items, and it was annoying to reorder them so often.

While the costs aren't insignificant compared with the costs of purchasing a laptop and maintaining a home broadband connection, smartphones are the cheapest form of Internet access.[13] Low-wage workers have been pulled into the smartphone market by these expanded options and lower prices.

However, the choices that low-wage workers make to buy smartphones instead of home broadband are also shaped by an important "push" factor. In the past decade, Internet service providers have been charged with excluding or deprioritizing low-income communities and communities of color in both rural and urban areas from infrastructural improvements that would have provided upgraded home Internet services on par with those offered to subscribers living in wealthier areas. This "digital redlining" results in inferior home Internet offerings to residents of lower income neighborhoods in cities across the United States.[14] These practices may serve to further push low-wage workers into using smartphones to secure their connectivity. In many cities, the legacies of these push and pull factors have sharply segmented Internet access by race, with neighborhoods with more Black and brown residents relying more on their smartphones for access than predominantly White neighborhoods.[15]

Smartphone companies have capitalized on low-wage workers who need to be connected and the gaps that exist in home broadband coverage. These companies not only offer cheaper, more limited plans but also subprime loans for expensive phones through third-party companies that, while charging exorbitant interest, make owning a smartphone *appear* affordable. Throughout my time in the store in DC, I noticed stacks of fliers advertising "leases" for new smartphones, offering to "get it now, pay later" (see Fig. 2.4).

Fig. 2.4. MetroPCS flier for SmartPay phone "leasing." Author's photo, June 2017.

This service, offered by a company called SmartPay, allows people to either lease or take out a loan to purchase a smartphone and related accessories with no credit check. In addition to traditional repayment with a credit or debit card, customers can also make cash payments at 7-Eleven, CVS, and Family Dollar stores. SmartPay partners with Straight Talk, which is the most widely sold prepaid carrier. These phones are sold at Walmart stores along with other brands that target low-income customers. In 2018, SmartPay had their Better Business Bureau accreditation revoked due to an overwhelming number of complaints and their failure to respond to them.

These leases have opaque terms, which often result in unexpected costs, making them akin to smartphone payday loans. Sales people in partnering mobile phone stores often advertise attractive terms, such as ninety days "same as cash," meaning the customer pays no fees or interest on their loan if they pay the full amount back within ninety days. Reporting on these services has shown that, in reality, many customers don't qualify for these terms and face shorter repayment windows and higher fees.[16] However, customers sometimes don't find out about the terms of their individual lease until after they've walked out of the store with their new phones. New York city sued T-Mobile for rampant sales abuses in their Metro by T-Mobile (formerly MetroPCS) stores that included enrolling customers into these exploitative financing plans without their knowledge.[17]

Being poor is expensive.[18] Like payday loans, these loans target those who can't afford them, without the cash on hand to pay the high up-front costs of buying a smartphone, and who may also have poorer credit histories that prevent them from qualifying for more traditional loans that offer better terms. While SmartPay likely evaluates all of its customers in the same ways, equal treatment in an unequal society still fosters inequality for low-income people of color. After decades of institutionalized discrimination in access to banks, mortgages, and other mainstream financial products, communities of color have been disproportionately targeted with marketing for high-cost financial services like SmartPay, along with subprime loans, prepaid debit cards, and check-cashing services. Terms like "predatory inclusion" and "reverse redlining" have been used to describe services like SmartPay that claim to democratize access to high-cost and necessary goods but have an ugly history of exploiting communities who have long been excluded from the financial system and lack access to credit, charging them higher prices for the same products that wealthier people get at a lower cost.[19] Services like these exploit those who lack access to traditional forms of credit to profit from essential goods, like phones, and reinforce a cycle of debt, default, and poor credit.

The mandate for low-wage workers to be connected isn't only leveraged in purchasing a phone but also in the phone plans they need to use those phones. James, a yoga instructor and sometimes handyman and mover, was paying $30 a month for a prepaid plan that entitled him to unlimited voice and text and a small amount of high-speed Internet access. "I don't do contracts unless it's a million-dollar contract," he told me with a laugh, "it's my most important bill, and I hate bills, so I like to keep it low. It's great, but when that high-speed goes out, whew! You're at AOL levels!" James often finds his handyman gigs on Facebook, where his past clients put him in touch with new people, and uses Facebook Messenger to send quotes and schedules. "Facebook just eats your data up. Once, I met my limit and Facebook just wouldn't even open, like, really? You need the Internet to just open the app? It's really slow." When this happens, James gets a text that alerts him his high-speed data limit has been reached and offers to "top up" his data for $10 for each additional gigabyte.[20]

Since 2007, low-income consumers have fueled the market for prepaid phone plans.[21] These are often the only plans available to people with low incomes, especially those with poor or minimal credit history. Prepaid phones are offered without annual contracts, meaning that customers pay for a predetermined amount of data, voice, and text messages, and deductions are made against this monthly allowance until the balance is gone, at which point the service is disconnected, the speed is slowed to a snail's pace, or a hefty fee is applied to continue service. Customers may top up their accounts at any time and usually pay fewer fees than those with annual contracts for switching providers or making changes to their plans. However, contracts are flexible and features and promotional pricing may end without much notice, resulting in unexpectedly high bills or service being shut off unexpectedly.[22] Prepaid users must also gather enough money to pay for their devices up front, which usually results in users opting for lower cost, lower quality phones with fewer features or high-interest phone "leasing" programs like the one described earlier. High costs and shifting and unclear terms of service presented problems to the grocery store stockers, bartenders, waitresses, and day laborers I interviewed, who often reported having their service shut off after missing payments for their handsets or after using their allowances more quickly than they had anticipated.[23]

Despite these devices' important role in enabling access for low-income users, people who rely on their phone for Internet are treated as "second-class citizens" online.[24] Prepaid users often experience lower quality services than wealthier postpaid clients. In 2017, the Federal Communications Commission reported that "post-paid users frequently are given priority over prepaid users on a given network during times of peak congestion."[25] This means that prepaid customers may experience slower download speeds than postpaid users simply because of the way they pay for their services.

At each point in the process of getting online, low-wage workers face a market that exploits their need to have phone and data services. The political economy of the consumer market for smartphones and plans makes it clear that the digital divide isn't an accident or the result of hurdles that workers struggle to scramble over because of their low incomes. Instead, the mandate that low-wage workers be connected to coordinate their lives is big business, and the market for wireless products and services has expanded to include them on exploitative terms. These terms are not only an issue at the point of purchase but also shape the strategies low-wage workers deploy to find and coordinate their work on a daily basis.

The Costs of Inclusion

Wireless companies aren't the only actors in the marketplace capitalizing on the mandate for low-wage workers to maintain their connectivity. Low-wage workers hustle for inclusion to work and rely on foraged Internet connections, commercial sources of free Wi-Fi, and free apps. In this process, low-wage workers enter a media ecosystem where advertisers and data brokers surveil and extract value from their data while offering them few protections for their privacy.

In addition to working for a hotel and restaurant, Damon also occasionally looked for cleaning gigs through job boards like Indeed.com. After chatting about all of the personal information sites like these require to create a profile and apply for a job—including education, location, email and phone numbers, and even sometimes a social security number to pass background checks—I asked him if he thinks about privacy while he's hunting for work online:

I've thought about that. I'm not concerned, but I just wonder because sometimes I would go on Indeed[.com] and then next thing I know I have a college calling me. It makes me wonder where does the information really go to?

Julia: Do you think they're sharing or selling your information?

Maybe. . . . I never thought about that. Do they do that? They sell people's information?

Julia: Sometimes!

Wow. I mean, Indeed[.com] has to make money. Yeah. That makes sense. I totally agree with that. . . . I definitely wouldn't have a problem with it because they're helping [me] get a job for free, so if they make money back by selling my info, I don't care as long as it's nothing crazy. I wouldn't care.

Julia: Okay. Have you ever gotten contacted by a business or anything that you think might be from that?

Yeah. I get calls from colleges all the time, and I haven't looked up information from them. One was called Post University. God, I mean, there's so many. I get a lot of random calls, now that you make me think of it . . . a lot of spam. . . . Sometimes they be calling about financial stuff, like help getting rid of credit card debt, stuff like that.

While Damon was initially surprised that Indeed.com may be selling his data, he wasn't upset about it and understood that he was trading his personal information for "free" access to jobs in his area. However, the examples he gave outline how targeted advertising can reinforce and even automate unequal treatment of vulnerable populations. Post University is a for-profit school that, like other for-profit schools, tends to spend large portions of their budget targeting low-income potential students with advertising emphasizing economic mobility while saddling students with debt.[26] Similarly, debt relief schemes often target people in economic trouble by offering to negotiate on their behalf with creditors and charging a large up-front fee for uncertain results.[27] By using free services, mobile apps, and unsecured, open Wi-Fi connections, low-wage workers secure access to crucial services, but within what Seeta Gangadharan calls a "privacy poor, surveillance rich" context that could also enroll them in predatory data tracking and profiling systems that may amplify their disadvantages.[28]

Estimates suggest that, in 2017, advertisers spent more money on ads targeting people accessing the Internet and social media platforms on mobile phones than they did on newspaper, magazine, movie, and outdoor advertising combined. It is expected such spending will outpace spending on TV ads in the near future.[29] This combination of digital advertising and disadvantage is the result of what Oscar Gandy calls the "panoptic sort."[30] Mobile Internet and app use, which is concentrated among low-income and communities of color who depend on it as their sole access to the Internet at higher rates than middle-class and White people, creates unique and sensitive data traces that have the potential to pinpoint individual users across devices, time, and space. Data brokers, including advertising companies, use these data traces, largely gleaned through social media sites, to place people into market segments that mark them as financially vulnerable.[31] Companies offering substandard services, scams, and exploitative financial services purchase consumer information from these segments to target low-wage consumers and other financially distressed groups with offers and advertising that may serve to concentrate their disadvantage.[32] The framing of the problem of low-wage workers' digital technology use within the digital divide obscures these new kinds of companies and intermediaries who are exploiting the expansion of mobile phone use among low-income consumers by collecting and selling data in exchange for free services.

The Invisible Labor of Connectivity

Within this exploitative political economy of connectivity, temporary disconnection from Internet and phone service isn't a surprise but a predictable feature of workers' everyday lives. As the next chapter explains, high-wage gig workers in professional fields who are saturated with email and ubiquitous connectivity find solace and renewal in breaks from their social media accounts and digital requests from colleagues, but this was not an option for many of the low-wage workers I interviewed. Low-wage workers employed a variety of strategies to get back online and compensate for disconnection from phone and Internet service. These strategies, from negotiating for extra time to pay bills to juggling multiple devices and relying on friends and family to take emergency calls, were a form of skilled and unpaid labor required to maintain connectivity in the midst of constraint. Workers did several different types of invisible labor to maintain communication when they weren't able to top up their prepaid plans or replace a broken phone right away.

Several interviewees mentioned interpersonal strategies they used to negotiate extra time to pay their phone bills in months when they couldn't make ends meet. Darnell, a manual worker living in Washington, DC, who works seasonally as a landscaper and mover as well as in snow removal under a variety of different bosses, explained:

> You know, I can't always pay that bill. Sometimes I'll just take a little and pay on it, but not be able to pay the whole thing [and it] just builds up. Sometimes I need to take that $80 and get some groceries. I'll have to wait to pay for it.
> Julia: How does it work when you don't pay?
> They'll shut it off, or they'll work with you, it depends on who you get when you go in [to the wireless store]. That happened about two weeks ago. I went in, but there was a bunch of new people working there and the manager was just watching them and they just shut it off and I had to borrow some money to get that back on. Sometimes it works, sometimes it don't, depends on who you get. It's crazy, it's like, damn! That's cold! It's not right to do someone like that.
> Julia: Did they shut it off right when you were in the store?
> It was like a minute after I left. So I was left out there with nothing, no phone. I had to tell my friend I was going to be late, so I had to find a way to call. A lot people, too, they feel funny about giving you their phone to use to make a call. I usually ask, but a lot of people don't like it. They think I'm gonna run off with it. That makes me mad, I mean. The whole process made me feel like it's a cold world, you know? Like nobody's gonna do nothin' for nobody unless they have to.

Darnell explained that he eventually found an acquaintance, who was waiting for his food outside a take-out restaurant close to the wireless store, who agreed to send a text to his friend to let him know Darnell would be late. Darnell has also committed to memory the phone numbers of his frequently called friends, family members, and work contacts, a necessary skill for someone with an old, low-end phone with an unreliable battery. Workers like Darnell mentioned negotiating with store employees, customer service representatives, and a variety of friends and family to smooth out the immediate disruption that disconnection caused in their daily routines.

Getting one's phone shut off is more common for those who depend on their phone for Internet access than those who don't.[33] This means that the people who depend on their phones the most are also the most likely to be without them due to financial pressures. As Mike's story at the beginning of this chapter illustrates, disconnection can also lead to lost income, creating a self-reinforcing cycle of disconnection and financial constraint that low-wage gig workers had to get creative to avoid.

Another strategy is to juggle multiple devices, often in various states of disrepair. Damon, whom I met during the promotional concert in the wireless store described earlier, explained the way he shuffled several smartphones to coordinate his work schedules. Damon and I coordinated our interview through his aunt, who texted me to confirm the time and place of his interview and told me to use her phone number if I needed to get in touch with him because "his phones are messed up . . . again." Damon came to our interview and apologized for the complicated message relaying, pulling out his phone to show me its smashed and inoperable screen. This was the backup phone he was using while saving up to purchase a replacement for his main phone, which after we chatted in the wireless store, had completely stopped working.[34] This was a huge problem as the bosses at his various gigs all relied on text messaging to schedule shifts and communicate about last-minute scheduling needs. As if on cue, while Damon was explaining the way he and his manager communicate about his weekly schedule, his phone lit up with a call from her:

When the schedule come out, my manager, she texts everybody the schedule.
Julia: Your personal schedule?
Yeah. She sends out like twenty-five texts. It takes her a while. And when I can't barely see my screen because it's all broken like this, it's a real problem you know? [His phone rings, he glances down at the screen.] Speaking of her, this is my manager right here.
Julia: [laughs] How can you even tell it's her?
I got a picture of the hotel as her picture, so when it comes up when she calls, I can see around the edges here, see? You can sort of see the background, that's how I know it's her. Excuse me [speaks on phone] Hello. What time is it? You think you can give me another thirty minutes? [Turns back to Julia] She said someone called out and she want me to come in early. . . . I don't wanna really end this right now. You think maybe it will last another thirty minutes?
Julia: We can do thirty minutes. That'd be great.
[Talks into phone] I'm in the middle of something. I'm gonna be getting on the bus around one. Is that alright? Okay. You can hold the shift down until

then? Okay. Cool. . . . Just give me an hour. Thirty minutes to get to the bus stop and thirty minutes to get to work. Is that cool? Okay, cool. That's no problem. Thank you. Bye-bye.

Julia: Does that happen a lot?

All the time. Today I'm actually doing laundry work. The laundry guy called out. So she calls me, usually calls me first, because she knows I want the extra shifts. But it means I gotta have a phone, you know? Otherwise, I just miss out on that. That why she has both my phone numbers. She always tries both.

Julia: She has both?

Well, we're very close.

While on paper Damon works a fixed number of hours that changes every two weeks, in practice he's on call, and even if he weren't, the way his manager distributes the schedule requires that he have some way to receive text messages.[35] To accomplish this, Damon enlists several phones in various states of disrepair as well as strategy in collaboration with his manager. He builds in redundancies to his digital hustle or layers of connections that he can call up in case one gets short-circuited.

While Mike and Darnell struggled to maintain their connections in urban Washington, DC, their experiences were ones of personal poverty surrounded by apparent public abundance; they were still able to find oases of Wi-Fi and were usually surrounded by people with working phones. For the low-wage workers I spoke with in rural Mainville, New York, this was not the case. Their experiences with being periodically disconnected from connectivity seemed less like looking for oases and more like scuba diving because they couldn't rely on foraged connections throughout their days. Instead, they had to plan for contingencies while they were without service until they could put together the funds they needed to top up their prepaid plans.

Ashley is a single mom who lives with her six-year-old daughter in two rooms in her grandmother's house in Mainville. Ashley moved in with her grandmother, who lives close to her daughter's school, after breaking up with her daughter's dad and being let go from a steady plastics manufacturing job a year before we spoke. She assured me that the situation is temporary, but she was struggling to save up enough for a rental through her part-time hours at a local fast-food restaurant and Walmart. She and I sat in her grandmother's living room, absentmindedly putting together a puzzle her daughter brought us to play with, chatting as her daughter offered us a continuous stream of other toys and books to get us to play. Ashley sighed heavily when I asked her if there had ever been a time when she was without a phone or Internet when she really needed it. She told me a story about the previous week, when her hours prevented her from either dropping off or picking up her daughter from school, requiring grandma to step in. Two weeks later, when her schedule changed again, Ashley was able to pick her daughter up from school and saw her teacher:

> She [the teacher] goes, "It looks like Audrey's feeling better!" and I'm like, what? She must've seen how confused I looked and said that Audrey was in the nurse's office with a fever sometime last week. They tried calling me during the day but got the message that my number had been shut off. They told my grandma when she picked her up, but her memory isn't always so good, and the fever was really low and the nurse said she was acting fine anyways, but still! Like, how do you not tell me that? I know it's my own fault, but I never really thought about that—they only had my cell number at the school, you know? I call into work to find out my schedule, so when I don't have minutes it's annoying, but I can just drive there and look at the schedule, you know? But how would I know something happened at school if I wasn't physically there? I went right to the front office and left a bunch of numbers for them, my numbers at work, my aunt's number, whatever, you know? Because I'm always running around between the two jobs, and my phone gets shut off all the time!

The school's administrative assistant and a nurse at the local pediatrician's office were also among my interviewees, and both women independently brought up this same issue during their interviews, saying it was frustrating to be unable to reach parents with requests or information about their children because their phones were temporarily shut off or they were "out of minutes" on their prepaid plans and couldn't call them back. After Ashley's interview, I followed up with these two women and they confirmed that having extra numbers in a child's file was becoming more common and helpful, but it often sent them on, as the school administrative assistant described it, a "wild-goose chase" to find the parent while trying not to divulge any sensitive information to any number of people along the way. While these issues affected the parents among my interviewees in urban areas as well, as Darnell pointed to earlier, they could usually rely on people who were hanging out in public or quasi-public spaces to relay messages or make phone calls. Rural residents, however, lived in towns without public gathering spaces and were often more socially isolated and living alone.

Mainville has an elevated unemployment rate and a poverty rate more than double the national average, and many of my interviewees there faced the same obstacles as the low-income Washingtonians in the study, struggling to save and pay for consistent wireless service let alone home broadband connections.[36] Mainville used to be home to several large manufacturing plants, but over the past fifteen years they slowly left town, leaving many of the town's residents unable to find the factory work that had sustained their families in the past.[37] Some of my interviewees in Mainville owned smartphones, others had feature phones, and many cycled through predictable periods of having their service shut off and being able to turn it back on again the next week. For most interviewees, these phones were their only phone line, so these periods of "predictable instability" were particularly frustrating.[38] As Ashley's story illustrates, disconnection didn't only affect their ability to earn income; it complicated other parts of their life as well. Being unable to afford their plans was an especially important problem for parents with young children, for whom getting cell phone service cut off meant losing their connection to their children's school and doctor.

This issue goes far beyond Mainville, New York. In the United States, more than half of all children live in homes with only wireless phone services, and this number is even higher for children living near the poverty line.[39] For some, workplaces, family, and friends provided supplementary connectivity when they weren't able to pay their bills or purchase more airtime. For people living in rural areas, without ready access to other public or semipublic infrastructures of connectivity like Wi-Fi hotspots in parks, libraries, and restaurants, being disconnected from phone and Internet services disrupted their everyday routines, affecting children and others who depend on daily care and communication.

Even for those low-wage workers who have relatively secure access, turning their phones off to have uninterrupted family time may still not be an option. Charlie, a contractor and mover in Washington, DC, whose phone played such an important role in his digital hustle, said that he "felt like a lost person" when he left it at home after rushing to reach a client's house on time. He dropped his head toward the table and shook it slowly as he explained, "I lost business that day because of that. There were four jobs that I lost that day because I left my phone. I achieved what I had to do that day, but I lost future business. It's like you run to take a step forward and trip over yourself and end up behind." For Charlie, constant connection was imperative to finding and securing work. Charlie's memories of lost income were a painful reminder of the simple act of forgetting his phone at home. His use of language suggests a view of the market as a race that he needs to be fast and connected to win. He can't afford a single misstep.

The digital hustle requires connectivity to phone and Internet services, so when cheaply made phones break, deceptively priced plans become too expensive, or unpredictable work hours don't generate enough income to top up airtime and data to stay in touch, low-wage workers perform invisible labor to compensate for disruptions to their connectivity. This labor is an expansion of the unpaid "work-for-labor" low-wage workers have to do to put themselves in the position to get paid.[40] Unlike the high-wage workers, who used their relatively higher incomes to outsource this work and purchase stable connectivity, as we'll see in the next chapter, low-wage workers had to accomplish this work on their own, adding to the already heavy burden of unpaid labor of their digital hustles and reinforcing inequalities in contingent labor markets.

Disciplining Maintenance

As this chapter has explained, the exploitative ecosystem within which low-wage workers secure their connectivity to phone and Internet services leads low-wage workers to do invisible labor to find access and compensate for lost connectivity, often by skirting the rules of fast-food restaurants and hanging out for extended time in public spaces. However, these strategies weren't without their own consequences. For some low-wage workers, especially Black men, foraging for Wi-Fi in public or semiprivate spaces threatened to bring them into contact with the police or other kinds of racialized violence. For these workers, not being able to afford their plans may have been what led them to seek out other kinds of access, but a lack of funds wasn't as salient in keeping them disconnected as the racism they experienced in public spaces.

At the time of our interview, Alex was working two part-time jobs while finishing his associate's degree at a small college in Washington, DC. He was living in Maryland and spent a lot of time commuting to work and school on public transportation, using his phone to keep him occupied. For his overnight shift receiving and unpacking food at a grocery store, Alex always made sure his phone was charged so he could listen to music during the long, lonely hours of unpacking boxes and walking up and down deserted grocery store aisles. Sometimes he'd stop at a Starbucks up the street from the grocery store and use an outlet there to charge his phone or download music and on-line readings for his courses in between his long commute and his shift. He recounted that, recently, a new manager at Starbucks had started cracking down on people sitting without buying anything. After seeing Alex a few nights in a row, the manager told Alex he had to buy something to use the outlet. When Alex pointed out several other people, including someone he thought was homeless, who were all doing the same thing, the manager told Alex he was going to tell them all the same thing. Alex insisted he would just be a few more minutes, but the manager didn't appreciate the "back talk" and threatened to call the police on him for loitering. Alex recounted his reaction, "I didn't need that before a long night, you know? I was mad, but she was right, I was sitting there without buying anything, so I don't think that deserved the cops getting called, but it wasn't worth it. So I just got out of there." Alex's experience isn't an isolated incident. It appears that chains that offer free Wi-Fi, like Dunkin' Donuts and McDonald's, have been more strictly enforcing their loitering policies to keep people from lingering in their stores to access the Internet.[41]

Damon, who'd recently discontinued home Internet service after being disqualified for a subsidy through his Internet provider, had been trying to avoid using up all the data on his monthly mobile plan by using free Wi-Fi outside the rental office of his apartment complex, which he described as "in the projects." He told me he didn't like standing outside for too long there because "they get on people for standing around, loitering, you know. Some people like me just be there to use the Wi-Fi but other people are trying to do bad stuff, so I understand." Like Alex, his complaint was colored with a shade of empathy and resignation for the perspective of the property managers. He said that while he didn't think it was right for management to call the cops on people using free Wi-Fi, he also didn't want people selling drugs or bothering his aunt as she came and went from their apartment and understood why they occasionally called the police for help. Nonprofits and other organizations have tried to set up Wi-Fi hotspots in other public spaces in his neighborhood, but being in public spaces in this neighborhood meant risking becoming ensnared in the violence, whether between gangs or with the police, that had recently claimed the lives of one of his good friends:

I hate it. I hate every second of it. Can't even live properly, like, it's not worth it to go outside to check my Instagram, you know? That can wait.

Julia: If you had to say what about it you hate . . .

[interrupts] The violence. Hands down. A friend of mine just got murdered in my neighborhood last month. [pauses] I'm used to it, it's nothing new, but I guess you can't really never get used to stuff like that.

Damon's and Alex's experiences illustrate the insufficiency of conceptualizing access as the only barrier to be overcome in creating equitable conditions for connectivity. Wi-Fi use in public or quasi-public spaces like fast-food restaurants or coffee chains usually comes with time limits of twenty to thirty minutes, a laughably short window to look or apply for a job online or fill out an application for unemployment benefits. Vagrancy and loitering laws have long provided a broad license for police and the public to enforce social boundaries about who is allowed in certain public spaces and disproportionately target Black and brown men.[42] When workers like Mike, Damon, and Alex are disconnected from the services they need, they face threats of criminal penalties for trying to secure access to Internet and phone services to find and coordinate work, and yet race is a largely unmarked aspect of public discourse and policies that address the digital divide. For some of the Black men in this study, low incomes may have been the most immediate cause of their disconnection, but their attempts to overcome this barrier saw them policed and harassed out of the public and commercial spaces that provided alternative sources of connectivity.

Conclusion

The low-wage workers in this chapter are not usually those that first come to mind when thinking about "wired" or "digital" jobs. Postindustrial theorists predicted that, spurred on by advances in information and communication technologies, the work of the mind would dominate the work of the hand, but history hasn't unfolded exactly along these lines.[43] The "network society" has indeed transformed the labor market, but instead of sweeping away low-level work, it has created new requirements for those in manual and service jobs.[44] For low-wage workers stitching together gigs, communication and connectivity are essential to securing paid work.

Far from being excluded from the digital economy, low-wage workers are expected to be connected. However, they're included on exploitative terms that reinforce existing inequalities rather than equalize opportunity. Phone companies, carriers, and digital advertisers exploit low-wage workers need to be connected by making connection appear affordable while offering hardware and services that break, invade their privacy, and create systematic challenges to maintaining connection. Since their connectivity is largely unsubsidized by employers and inadequately supported by federal subsidies, this connectivity mandate represents a new tax on low-wage workers' stagnating incomes.

To compensate for routine disruptions, low-wage workers perform invisible labor that makes use of interpersonal, technical, and logistical skills. From juggling multiple devices, relying on family and friends to relay messages, or foraging for Wi-Fi in public or commercial spaces, low-wage workers did unpaid work to maintain their connectivity. However, not all were able to rely on these strategies in the same way. For some workers, especially Black men, access to public and commercial spaces was foreclosed through racist stereotypes and the threat of police and other violence.

All too often, the ways we talk about digital technologies and labor market inequalities focus on the many ways that social groups are excluded from the benefits of digital technology use, such as gaining technology skills that would earn them higher wages. However, as this chapter illustrates, it's becoming just as important to our understandings of inequality to examine the terms on which people are included in the digital economy. Focusing on the mechanisms of exclusion has obscured the profits and penalties that confront low-wage workers as they try to maintain connectivity. Continuing to frame these issues as coming from a lack of resources, skills, or motivations obscures how the exploitative terms on which they've been included serve to reinforce, rather than alleviate social inequalities.

The problems that low-wage workers face are rooted in the fact that connectivity to mobile phones and the Internet is now essential to their ability to participate in insecure labor markets, and "exploitation thrives when it comes to the essentials."[45] The necessity of connectivity has made low-wage consumers a profitable market segment for many wireless companies and has generated a predatory subprime loan that makes owning a smartphone appear affordable to those with poor credit and low incomes. When the bills pile up and low-wage workers turn to foraged connectivity and free apps, they face penalties—from getting a ticket for loitering to being targeted as financially vulnerable—that accompany using their phones as their sole connection to the Internet.

Access to digital technologies and the Internet alone cannot promote equal access to entrepreneurial opportunities. To be sure, securing reliable, free or low-cost access to the Internet and mobile phone services for low-income people is an important first step, but this doesn't solve problems of equity in access to economic opportunity. As Jennifer Schradie explains, "the individualized Internet cannot combat institutional marginalization."[46] Acknowledging that connectivity is required for navigating government, labor market, education, and health care today—and is therefore a public good—entails the consideration of a host of other structural conditions that are the result of public choices that have, on the one hand, required workers to be connected, and, on the other, left them completely unsupported in getting online. Leaving connectivity up to the market has gone a long way to getting more people connected to the information and phone services they once couldn't afford, but it hasn't ensured equity. On the contrary, it has worked to further entrench many existing inequalities.

3

Comparative Advantages

Digital Privilege and the Ease of High-Wage Hustling

The first time Salena and I tried to meet for a video-conferenced interview, her smiling face appeared on my laptop screen from what appeared to be the inside of a car. Salena apologized for the awkward location, explaining that, after we had scheduled the interview the previous week, a new project dropped into her lap, and she got a surprise visit from her parents. She was Skyping me from the passenger seat on a road trip to Lake Tahoe, balancing her laptop on her knees on the drive up into the mountains, attempting to get some work done on the project. "Sorry it took me a few minutes to get connected!" she apologized, speaking into a headset, "I tethered my laptop to my phone for Internet, and the signal isn't great up here." As she spoke, Salena's mom leaned into the frame from the back seat of the car and waved. We all laughed a little at the circumstances, and Salena said that she'd have more free time after the new year.

Before our next attempt, a few weeks into the new year, I reviewed my fieldnotes from my first glimpse into Salena's life; I was looking forward to asking her about the many different kinds of labor that made up her digital hustle. Salena seemed like she was doing well as a public relations consultant and had the resources she needed to do it. She had skills and talents that were in demand by her clients, was well-connected to the Internet through multiple digital technologies, and had a personal life that included impromptu vacations. After months of interviewing low-wage workers striving for full-time work, Salena's well-resourced flexibility seemed like a gig worker's dream come true. However, during our interview, she explained that what had looked like success to me felt much less glamorous to her, saying, "It's basically my New Year's resolution to work less. When my parents came to visit, on that trip actually, I realized it'd just become this thing where I was online all the time."

Left to Our Own Devices. Julia Ticona, Oxford University Press. © Oxford University Press 2022.
DOI: 10.1093/oso/9780197631288.003.0004

In sharp contrast to the low-wage workers in the previous chapter, Salena wasn't scrambling to overcome barriers to connectivity; instead, she felt overly connected to her digital technologies and work. She mentioned that, during her vacation, she would head to the hotel bar to log into work while her parents watched movies in their room and hang back in their room while they headed out for a meal in the restaurant. While she was articulating her frustration with herself for allowing this to happen, I was struck by something else; Salena's sense of being inundated with connectivity to work felt as if it was happening in a parallel universe to the one workers in the last chapter were navigating. Unlike Mike's daily sojourns to McDonald's to access both the Internet and work, Salena had, in her mind, too much of both. Her problems were not ones of scarcity but abundance. If low-wage workers are navigating in a drought of connectivity, high-wage workers are navigating in a flood. Water is essential to life in both cases, but workers at either end of the labor market face very different kinds of problems with it.

Salena leveraged several different kinds of "capital," or symbolic, cultural, social, and economic resources that allow her to fit into the world of high-wage consulting, to politely reschedule our first interview. From her smartphone, laptop, and data plan that provided her Internet connection to her knowledge of how to connect her laptop to her smartphone, and an apologetic demeanor, Salena was able to use many kinds of capital to attempt to fit her work alongside her life and other obligations. As social theorist Pierre Bourdieu explained, these different kinds of capital allow people to appear to fit naturally into dominant social groups and navigate social institutions.[1]

Some scholars refer to the knowledge Salena used as digital "skills." But in the hands of workers at either end of a polarized labor market, the same skills are often received in very different ways. For example, while Salena's knowledge of where to find Wi-Fi in her hotel and her ability to connect to it was unremarkable to her and probably also the hotel staff, as we saw in the last chapter, Alex's same skills of connecting to Wi-Fi at a Starbucks near his grocery store job got him kicked out and threatened with arrest. While calling these "skills" can make us think about this knowledge as an individual accomplishment or deficit, the term "capital" resocializes this knowledge into a field that includes institutions and power and results in some workers being able to leverage their skills for opportunity while others are punished for them.

Although workers I interviewed used different kinds of capital, the two types of capital that were most important to them were economic and social.

Economic capital refers to money, or resources that can be directly converted into money, while social capital is made up of the social connections one has that can be effectively put to use, whether that's a free drink from a bartender one knows or learning about a job opening from a relative.[2] The field of high-wage gig work sanctioned high-wage workers' capital, translating it into economic opportunity and ubiquitous connectivity.

Using interviews with high-wage gig workers, this chapter takes a deeper look at the social conditions that have facilitated this flood. Highly educated white-collar workers are thought to possess superior skills, knowledge, and access to technologies that enable their smoother and more successful digital hustles. However, in this chapter, I'll point out the ways institutions of high-wage gig work *allow* them to exercise these skills. The ways the political economy of connectivity, social media, as well as their colleagues, managers, and clients expected and rewarded their connectivity produced digital privilege for these workers, making their smooth hustles look like individual achievements. This chapter focuses on my interviews with high-wage workers but also draws comparisons between their experiences and those of low-wage workers to highlight the role of digital privilege in reproducing the inequalities between them.

The White-Collar Frame

Salena's context of abundant Internet connections, mobile data, and communication with clients is well represented in our scholarly and popular understandings of digital information and communication technologies. So far, we have largely understood the consequences of digital technologies in the world of work through the experiences of highly connected white-collar workers like her. Workers like Salena have been treated as the unmarked, standard population through which social scientists examine the issues that result from the increasingly mandatory use of smartphones, laptops, email, social media, and other Internet-connected applications and devices in the workplace, while low-wage workers, where they appear in this literature at all, are understood as anomalies. [3] Indeed, a search through any stock image website for the terms "digital technologies" and "work" is sure to yield pages of photos of white-collar workers using laptops, tablets, and smartphones in sundrenched offices and coffeeshops, and very few, if any, representations of workers wearing uniforms, working in cars, hospitals, or restaurants. Rather than seeing the low-wage workers in the previous chapter as "outsiders" to the otherwise unremarkable digital saturation of high-wage work, in this chapter, I flip that script to look at the experiences of high-wage workers like Salena as remarkable instead of familiar.[4]

Media scholars have used the idea of "frames" to analyze the ways institutions of cultural transmission, like news media, can place information within a field of meaning that guides audiences' interpretation in particular directions while making other possibilities more difficult to see. For example, Joe Feagin has pointed out a "white racial frame" that pervades everyday life and universalizes a uniquely White experience of the social world as normal and unremarkable.[5] In this same spirit, our understanding of digital technologies and work in the United States has been characterized by a white-collar frame. To make this frame visible, this chapter moves the experiences of white-collar workers away from the center of the frame to highlight not how low-wage workers are uniquely constrained and exceptional but how high-wage workers are the beneficiaries of institutions that sanction their capital in ways that produce digital privilege.[6]

Digital Privilege

Digital privilege is a "package of unearned assets" that high-wage workers can rely on to maintain ubiquitous and stable connectivity to high-quality Internet connections but about which they are mostly unaware.[7] These assets include issues usually referred to under the "first-level" digital divide such as access to physical technologies and being able to afford premium Internet and data subscriptions, and "second-level" divides in skills, activities, and motivations.[8] Scholarship on these divides has used Bourdieu's idea of capital to examine how social and cultural dynamics pattern both participation in and exclusion from digital spaces.[9] Scholars theorize that technological skills are a kind of capital, and as such, they can exacerbate inequality because some groups use them for "capital-enhancing activities" like visiting a museum online while others use them more for entertainment and socializing.

However, the same skills in the hands of individuals from different social groups can be received in very different ways by different fields or institutional arrangements.[10] Capital doesn't exist in a vacuum; it becomes valuable to the extent it's legitimized by social institutions like workplaces or schools. Institutions exist within fields that influence what kinds of capital get recognized as valuable. The ways that fields legitimize some types of capital produce privilege for some and marginalize others. Digital privilege is produced when the field of high-wage gig work, which includes the workplaces, social media sites, and even markets for the consumer goods high-wage workers navigate to do their work, legitimizes their economic and social capital and their use of digital technologies and skills so they appear as individual accomplishments rather than the result of that very capital.

Despite the importance of examining the role of privilege in the reproduction and endurance of social inequalities, we don't have many studies that examine the role of digital technologies in the reproduction of privilege. Privilege reproduces inequality "behind our backs"; it requires no effort or intention on the part of the people who benefit from it. Identifying digital privilege allows us to point out the invisible structures that make connectivity feel normal and unexceptional for some but not for others.

Though high-wage workers have their fair share of problems with digital technologies in their everyday working lives, I use the term "privilege" to draw attention to how these problems operate against a background of systematic advantages that becomes invisible in both scholarly and wider public conversations about things like "overwork," "burnout," and "work-life balance."[11] For these problems to emerge as problems in the first place requires a connectivity to the Internet and to colleagues that is unproblematic, ubiquitous, and unremarkable; it requires workers that have the money and skills to buy and use ever more sophisticated technologies, software, and associated services; and it requires that their constant use is authorized and rewarded in their workplaces, not punished. In this light, studies about the problems of white-collar overconnectivity illuminate the ways particular and contingent kinds of digital privilege can disappear into the background of our everyday working lives.

However, privilege is relative and white-collar jobs haven't escaped the effects of rising insecurity. Digital privilege is also intertwined with other kinds of privilege, such as racial, gender, and class privilege. The careers of high-wage workers like Salena illustrate how jobs have changed as a result of neoliberal globalization, with decreasing job tenure and organizations' increasing use of independent contractors instead of full-time employees.[12] In fact, many of the workers I describe in this chapter as high-wage would not, by many accounts, be thought of as earning a high income. Some of the workers in this chapter became independent workers because of layoffs or a lack of salary growth in their previous positions. Some of them held down full-time white-collar jobs while also starting their own small businesses. Others did a significant number of independent contracting gigs to fill in gaps between their salaries and the rising cost of living or a child's college tuition.[13] However, these workers all worked white-collar jobs and earned significantly more than the median income in their cities and, relative to the low-wage workers in the last chapter, enjoyed a comfortable level of financial security.

In Chapter 1, I described how both high- and low-wage workers relied on different kinds of labor for their digital hustles as they worked to cultivate and manage their various streams of income. I illustrated how the edicts of the culture of entrepreneurial capitalism aren't confined to white-collar workers and can be found in parts of the labor market we might not have expected. In this chapter, I'll explore the privileges of hustling with a white collar and how the field of high-wage gig work sustains theses workers' successful digital hustles.

Economic Capital and Naturalizing Digital Privilege

In the previous chapter, I explained that the political economy of connectivity to phones, data, and the Internet exploits low-wage workers' need to be connected. While employers had a de facto requirement for low-wage workers to have phones and Internet access, they didn't provide any support for their connectivity and sometimes explicitly punished or forbade them from having their phones at work. Workers told me how they hid their phones under uniform tops and in aprons and checked their messages in bathroom stalls, break rooms, and nearby coffeeshops. By contrast, in this chapter, I'll illustrate how this same political economy works to the advantage of high-wage digital hustlers, who use their capital to reinforce and strengthen their infrastructures of connectivity as well as their ability to compete and build their businesses.

The high-wage workers I interviewed were awash in Internet connectivity. Across their homes, cars, and offices, they routinely relied on multiple mobile devices, including smartphones, laptops, and iPads, robust access to high-speed Internet, and often, a dedicated staff to help troubleshoot technical issues. Penelope, a lawyer with her own small firm, pulled two nearly identical phones out of her bag at the beginning of our interview and asked, sheepishly, "Is it going to be really obnoxious if I just have my phones out? I'd love to be able to put them away and just focus totally on this . . . and I don't want to skew your data or anything. . . . I'm sorry! I'm waiting on a few calls." She admitted to feeling like having two phones on her "all the time" was "a little much," but explained that "I'm on call pretty much 24/7, so it's nice to have the separation between the personal and work phone," adding that "if I had client information on my personal phone and left it in a taxi, it could be a breach of client confidentiality, so I don't like to carry it around when I don't have to."

Penelope explained that she had just remodeled her home office space and contracted with an IT company that provides installation, security, and data management services for small businesses like hers. After working at a large corporate firm for several years, she was overwhelmed with everything she didn't know about setting up her own Internet, data storage, and security, and asked around to find a company she could hire to take the burden off her own shoulders. Once she had hired and begun to rely on them, she was a little embarrassed to admit how often she contacted them for help with small technical roadblocks, "Yesterday, I called them [the IT company] because I locked myself out of my own Google Drive!" Much like the employers of low-wage workers in the previous chapter, Penelope's clients expected to be able to reach her at all times, but unlike them, she could pay for the services, technologies, and staff needed to maintain continuous access and solve her routine problems.

On the surface, Penelope was in a similar situation to Damon, the low-wage worker in the last chapter who also juggled multiple phones and had to answer a phone call from a manager in the middle of our interview. Like Penelope, Damon was expected to be reachable so his managers and clients could call him into his shifts. However, unlike Damon, who was using his own old phone (despite its broken screen) and relying on his aunt as another point of contact because his main phone, which was also inherited from a friend, was broken, Penelope chose to purchase separate phones because of personal preferences for segmenting her work and personal use, and to protect her clients' information. When Penelope ran up against a small technical roadblock, she had the resources to put someone else's technical skills to work for her; when Damon's phone became inoperable because of what he thought was a virus, his only option was to find a different phone to use. While the political economy of connectivity weakened Damon's infrastructure, it strengthened Penelope's. She was able to leverage her economic capital to customize her work technologies to suit her personal preferences and make her work easier.

Near the end of my interviews, I'd often casually ask which phone and Internet carriers participants used, whether they liked them, and their typical monthly costs. While the large majority of low-wage workers quickly rattled off the costs associated with their phone service and data plan, the high-wage workers often admitted they "weren't one-hundred percent sure" of their exact costs or that they hadn't "checked the bill for a while" because they were enrolled in an auto-pay service.[14] The indifference of high-wage workers to their monthly bills shows that although the ability to connect to the Internet and use digital technologies has become ubiquitous in many kinds of work, the costs of this connectivity are only naturalized within the context of white-collar gig work.

Geoffrey Bowker and Susan Leigh Star point out that daily activities in "communities of practice" can naturalize certain objects, skills, or knowledge, allowing members of that community to forget the effort that goes into establishing or maintaining the thing in the first place. Using the example of electricity, Bowker and Star point out that we don't consider the science, technology, and labor that goes into powering our homes. "We no longer think about the miracle of plugging a light into a socket and obtaining illumination," in large part because electricity is easy; it's simply built into our daily lives.[15] The field of gig work naturalizes the capital necessary for ubiquitous connectivity and makes it harder to see the money and labor that it takes to make it possible.

However, for those who struggle to pay their electricity bills and are often cut off, flipping a light switch and seeing the lights flicker on might be a memorable moment. As Shamus Khan points out, *ease* is a key indicator of privilege. In studying students at an elite prep school, Khan noticed that it wasn't the possession of rarified tastes in classical music or art that allowed students at the elite school to become elites themselves but rather feelings of comfort or familiarity with elite practices like formal dining.[16] The ease with which high-wage workers executed their digital hustles was not only the result of their individual skills but also the result of indifference bred from familiarity.

Social Media and Social Capital

High-wage workers' digital hustles were also helped by other social sites of capital, like social media. For both high- and low-wage digital hustlers, social media was an important online context for cultivating new business and keeping in touch with previous clients. However, sites like Facebook, Instagram, and Twitter served to exacerbate existing inequalities in the social capital of these two sets of workers. By connecting them to their offline social networks, social media sites facilitated and rewarded the social capital of high-wage workers at the same time they constrained and cut off access to work for the low-wage workers.

High-wage freelancers find work in many different ways, including through staffing agencies, professional networks, personal referrals, and online marketplaces like Upwork, Fiverr, and PeoplePerHour.[17] Regardless of the method, high-wage workers felt that a professional and active online presence was important to finding work, especially toward the beginning of their careers. Some workers recalled stories of investing hours into their social media profiles and content to "build their brand," reaching out to friends and former colleagues to update them on their new ventures; while others, like Amy, a freelance designer in suburban Rancho Rio, California, took a more relaxed approach. When I asked her how she typically got new gigs, she replied:

> I basically get work word of mouth . . . when I started out, and was actively scoping out jobs. . . . I used the websites that are for freelancers to scope out gigs. . . . I've never actually gotten a gig through them, but I've put my portfolio on there. There's a ton of websites. . . but yeah, all the projects that I've had have been through like friends of friends.
> Julia: What nice friends!
> Yeah! I make friends a lot. . . . I know a lot of people. And also, like, people have just come out of the woodwork once they found out what I was doing. . . . People who I have fallen out of touch with reach out to me being like "Oh you're doing design? Send me your portfolio because we need someone for this project."
> Julia: How do they find out about what you're doing?
> I'm pretty active on Instagram, Facebook, Twitter. Most of the stuff I post isn't about work, but random stuff I'm into. But sometimes I'll post a photo of something I'm working on, or if I have a question about something work-related. So I guess that's how people see it. Some people are intense about promoting themselves, and that's not my style. I just kind of post about whatever I'm doing and things find me.

While Amy's explanation suggests that work just "finds" her without too much effort, in reality, Amy's social networks, which she'd cultivated from several years of design work in and around San Francisco–based start-ups, brought work to her. The ease of her social media practices and naturalness of finding work through her social networks obscures the "hard systematic work" of learning the rules and norms of these media that came before.[18] She distances herself from the kind of promotion she's seen others do online, perhaps because she hasn't needed to.[19] Her social media practices merge her professional and friendship networks, helpfully blurring boundaries between friends and clients to help her source new gigs.

Social capital has been defined in many different ways, but it usually refers to networks of interpersonal connections that can be put to use effectively. Social capital plays a large role in shaping our access to information, goods, and even economic mobility.[20] Early scholarship on social media was concerned that it may lead to a decline of social capital. Since then, scholars have demonstrated that the Internet and social media are beneficial for the maintenance of social capital.[21] But not all users experience these benefits equally. It's not the mere fact of Internet or social media use but how people use them that shapes the effects of technology on social capital.[22] Existing patterns of social inequality— such as income and education—shape people's online activities. For example, Hargittai and Hinant found that wealthier and more educated people used the Internet in ways that they presumed would benefit their accumulation of social capital.[23] This stream of research has focused on the important role of digital skills in facilitating the use of the Internet in "capital-enhancing" ways, pointing out that long-standing patterns of inequality shape which users have access to which skills in the first place. However, the Internet is not only a place where people do different kinds of activities requiring different kinds of skills, the Internet and social media platforms are part of the fields of precarious work, producing and altering social inequalities in the process.[24]

Unlike Charlie, whom we met in Chapter 2, Amy didn't need to manage her reputation obsessively through labor platforms, or send up flares about the kind of work she could do, hoping to reach strangers in need of a designer. Her potential clients were already in her network of Facebook friends. The differences between Charlie's and Amy's approaches to their online reputations can be explained through what communication scholars call diversification theory: the tendency of marginalized groups to use communication technologies, like social media, to build bridges beyond their own social networks and fill in gaps in their own social capital, and the parallel tendency of more powerful groups to use these same technologies to reinforce their already capital-rich in-group networks.[25]

Amy's normal everyday use of social media allowed her to translate the social capital of her online networks into the economic capital of new clients. Charlie had to use Craigslist and many other platforms to get connected with businesses and people who needed his services because those people weren't already in his social networks. In this light, Amy's approach to her digital hustle looks more casual and natural because her social networks and years of practice at the idioms of social media had already set her up for success. Amy's social capital is naturalized within the context of her social media use; she doesn't have to think about how to strategically use social media to cultivate gigs—it just happens. Workers like Charlie, however, work very hard to use technologies to expand their networks and overcome their lack of capital.

"Poaching" in the Office

High-wage workers' social capital didn't only allow them to use their digital technologies to get paid, it also allowed them to intentionally blur the boundaries between professional and personal uses of digital technologies. While many of the independent workers in this chapter didn't have regular offices, or worked from home, some did regularly report into an office as contracted workers. These physical workplaces facilitated their digital hustles because their use of digital technologies was seen as a core part of their work. By contrast, the low-wage workers were often explicitly forbidden and punished for their on-the-job use of digital technologies, even when it was directly related to their jobs, because managers and other authorities defined their use as outside of their scope of work.

Grace, a government contractor working for a federal agency, told me about how she usually avoids doing any personal Internet browsing at work because of strict government rules about using work computers for personal purposes. She assumes everything she does is tracked. But she told me about an important exception to this rule: "I try to do as much personal stuff as I can on my phone, but I will use my desktop to go through my LinkedIn contacts, see where they work, look at the job postings from where they work, just to see if anything looks interesting to me and reach out. I do it on the computer because I'm old and I can't see the text and it's easier to search." She said she figures if anyone ever sees her doing that, or asks about it, she can say she was making connections for the project she's currently working on. She compared her infraction to that of a coworker's to put it in context,

"We're all so underutilized on this project so everyone is always on YouTube and Facebook at work, so what I'm doing is at least professional . . . there was a guy who left a fax that made it pretty clear he was using a government fax machine to run his side business . . . that was a whole situation." Grace's workplace, and the surveillance she experienced there, sanctioned certain kinds of digital technology use and not others. Grace was able to get away with using LinkedIn because it's conceivable this could be a normal part of her daily work tasks. Michel de Certeau calls this work that workers do on their own behalf while appearing to work on behalf of an employer "poaching."[26] Grace was able to "poach" in the course of her daily work because browsing LinkedIn looked like something she should be doing. In this way, white-collar workplaces helped high-wage hustlers use technology to translate their social capital into economic capital.

This sly use of "company time" was a more difficult, and often impossible, task for low-wage digital hustlers. In fact, many of the low-wage workers shared stories about being explicitly forbidden and punished by their employers for using their digital technologies. Lannie, an artist who once worked as a manager at a Starbucks, told me that she had gotten so frustrated with the workers at the store using their phones during their shifts that she instituted a policy that if she saw them using their phones behind the counter, she would lock them up in the safe along with the cash and only give them back at the end of their shift.

Veronica, in her first few days in sales at a retail technology store, recalled being advised to use her own personal smartphone instead of the store's one slow and outdated desktop computer to look up information about products for customers. She was excited about the signal she thought this sent, explaining, "I'm a nerd about this stuff, so I was happy to use my own phone, it's better than the computer anyways . . . this is a cool company . . . they're letting us use our own phones." But, a few days later, upon seeing her and her coworkers passing time on their phones behind the counter while a customer was browsing the shelves, Veronica recalled that her manager had some sharp words for them during their breaks:

> She really got serious about it, you know? . . . Like, on each of our breaks . . . chewed us out for being on our phones with the customer there. But she didn't see that we already asked if they needed help! It's not like we're ignoring them. I was actually filling out all my new employee paper-work on my phone, like, really?! How am I gonna fill this out? Using that old

computer up there that'll take a million years? You don't even ask, you just assume we're screwing around and got mad.

The low-wage workers reported that although some of their jobs depended on their digital skills and their personal phones, their managers and employers still didn't define their use of these technologies as a part of their job; therefore, their use wasn't seen as a valued skill by those in power. Instead of being able to "poach" time to continue their hustles while in the office, as the high-wage workers did, low-wage workers were disciplined for what their high-wage compatriots took as natural parts of their jobs—and had their devices locked away in safes or strictly controlled.

The fields of high- and low-wage precarious work translate workers' social capital in ways that benefit high-wage workers and put low-wage workers at a disadvantage. In addition, the contexts of high-wage workplaces define the technology use of high-wage workers as important parts of their jobs and therefore allowable within the workday. For low-wage workers, even when they're expected to use digital technologies in the course of their work, their use isn't defined as central to their jobs and is therefore strictly controlled by managers. In his study of teachers' approaches to students' playing with digital technologies, Matt Rafalow found that the activities that get defined as digital skills aren't universal across settings but deeply shaped by social inequalities. While teachers in an affluent private school encouraged their students to tinker and play, and defined this play as a valuable digital "skill" that would prepare them for the labor market, teachers in lower income public schools disciplined their students' digital play, defining these skills not as skills at all but as a distraction from learning the real skills students would need to enter the workplace. He explains that "for skills to have meaning as valuable in a given institutional context, they must be validated by the context in which they are deployed."[27] For the teachers Rafalow observed, the class and racial context of their students and schools shaped the ways they translated—or didn't—students' play into valued skills.

High-wage gig workers were able to "poach" time to work on their digital hustles at work because their workplaces, managers, and colleagues understood digital technologies to be central to their jobs. Even though it was just as important to low-wage workers, this same activity was defined as distracting and unimportant. The definition of what counts as a digital "skill" is contingent on whose hands the phone or laptop is in and the kind of places they work.

Conclusion: The Politics of Digital Privilege

White-collar freelancers appear to have come out on top of large-scale changes that have shifted risks away from organizations and toward individual workers and carved out ever-smaller spheres of obligation between employers and employees. In their theories about these shifts, social commentators have pointed to this group—highly educated, white-collar knowledge and creative workers—as an indicator of the ways digital technologies will inevitably shape the world of work for the rest of us.[28] However, the emblematic status of highly connected knowledge work in the networked economy has obscured the contexts that have paved the way for its ascendance. This lack of context in literatures addressing the connection between digital information and communication technologies and work, as well as their consequences for workers, has led to an unacknowledged white-collar frame in our understandings of digital technologies and work. While the previous chapter showed how low-wage workers and their practices are made invisible by this frame, this chapter denaturalizes the frame itself and describe the contingent social structures and contexts that afford it.

Narratives about digital divides and skills gaps frame our assumptions about the ways digital inequalities work in society. Social stratification based on differences in education and income do shape the kinds of things people do with their technologies. However, these frameworks focus our attention on individuals and the groups they come from instead of the social contexts where they use their technologies to live their lives. In Chapter 2, I explained how research on gaps in digital skills is often understood as pointing the finger at low-income populations as deficient. While recognizing social patterns in the kinds of knowledge and activities that people engage in around digital technologies is important, "skills" thinking doesn't give us the tools to interrogate institutional arrangements or fields of precarious work.

The knowledge and practices that eventually get defined as digital skills are better understood as the end result of political projects rather than objective measures of job-readiness or ability.[29] In this chapter, I've shown how the digital skills that are seen as central to white-collar digital hustlers' work are defined and legitimated by the field of high-wage gig work, not by the individual. The kind of workplace workers find themselves in goes a long way to define what constitutes a particular activity as a valued skill or, conversely, as a time-wasting or even fireable offense.

Both low- and high-wage contingent workers use their digital hustles to make a living. They rely on their digital technologies to find work, communicate with clients, and execute many of their daily tasks. However, as this chapter illustrates, they do so with sharply different sets of resources at their disposals. The privileges of their higher status, personal networks, work contexts, and incomes allowed high-wage workers to create and execute their hustles to ensure constant connectivity and limit their technical struggles.

Comparing the experiences of high- and low-wage workers makes it easier to see the advantages that make high-wage workers' hustles appear more seamless and their skills seem more valuable and allow them to disconnect and cope with their hustles in ways that aren't available to their lower wage compatriots described in the previous chapter. The field of high-wage gig work sanctioned these workers' capital to produce digital privilege, which made their smooth digital hustles look like the accomplishment of individual ability and skill instead of systematic forms of privilege. Digital privilege, like other kinds of social privilege, operates in the background and takes no effort on the part of the privileged to maintain.

Digital privilege becomes naturalized against the exceptional conditions of high-wage white-collar work. Whether in physical offices or in online work contexts, digital technology use is seen as central to high-wage work, and these contexts facilitate and sanction the digital hustles of high-wage workers. At best, digital technologies are seen as marginal to low-wage work; at worst, they are explicitly forbidden by managers. From this vantage point, it becomes clear that the experiences of the low-wage workers in the last chapter aren't only the result of individual constraint or lack of power but are also linked to the privilege of the relatively more powerful workers in this chapter. We need to better understand what kinds of uses of digital technologies get defined as skillful or skilled in particular contexts, which means understanding digital skills as social, political, and cultural constructions rather than as requirements or statements of fact.

Digital privilege constructs digital inequalities just as much as constraint does. While research on digital inequalities has documented the myriad ways marginalized technology users are constrained by classist, racist, and gendered assumptions on the part of designers and by lack of income, education, or access, we have been slower to acknowledge the ways that constraint is only one-half of the equation of inequality. Examining the conditions that foster digital privilege opens up space for a broader structural critique of the conditions that foster digital inequalities. It can often seem like digital inequality is something that only affects some. In reality, it affects us all. It is necessary to look for the ways that power and privilege produce digital inequalities as well; in fact, as I'll explain in the next chapter, it's impossible to see one without the other.

The invisibility of digital privilege is political, meaning that it's a feature, not a bug, of the power it relies on to continue operating in the background of some of our working lives. The invisibility of this privilege is political because it has consequences for who we see as belonging within the community of digital workers and who gets excluded. As I'll explain in more detail in the next chapter, when it comes to digital technologies and work, we have plenty of policies about inclusion; what we don't have is a good grasp on the language to talk about its politics.

4

Suspending the Hustle

Diverging Strategies of Resistance

Rae came to California from Chicago to be a professional dancer, but for now, she was living with family outside of San Francisco to work and save up enough money to move to Los Angeles. I met her while she was waiting around for a screen repair at her phone store, where she also came to pay her bills in cash. When I asked her if she used her phone for anything work-related, she could hardly stifle a laugh. "It keeps me from quitting!" During our interview a few days later, she explained that, on her first day working as a bartender at a casual dining restaurant, she was given a list of corporate rules that instructed her to keep her phone in a locker in the back room, which Rae dutifully followed for exactly one day. Rae was already annoyed about having to take this job, and this rule made it that much worse. Waiting around for the after-work happy hour crowd without much to do was boring, and she was actively looking for gigs as a dance and fitness instructor and wanted to be able to monitor email and phone calls for potential leads. After seeing the other bartenders, servers, and even managers kept their phones in aprons, or pants pockets, Rae decided to join them. She told me that being able to use her phone was one of the few things that helped stave off the boredom and keep her from quitting: "I can't come and go as I want. I have to be there when they tell me to be there, I can't go anywhere all shift, I have to be in the same place . . . I have to smile and laugh and everything . . . pay attention to every little thing. . . . If I couldn't play games on my phone, I would've quit the first week."

To reclaim some control over the conditions of her work, Rae broke the restaurant's official rules and used her phone to play games behind the bar. As she smiled, laughed, and "pay[ed] attention to every little thing" for her customers, playing cell phone games gave her a chance to momentarily suspend her performance as an attentive server. However, it was common knowledge that supervisors would only tolerate this small violation, "as long as you're not on it [cell phone] when there's customers around." Rae was

Left to Our Own Devices. Julia Ticona, Oxford University Press. © Oxford University Press 2022.
DOI: 10.1093/oso/9780190691288.003.0005

allowed to use her phone to escape the demands of her job, as long as there weren't any customers around to see this break in her performance.

Rae's use of her phone to temporarily escape from her job was an act of resistance against the expectation that she would devote her undivided attention and emotional energy to doing her job. In jobs in food, hospitality, and across many different service industries, workers' bodies, words, and even feelings are pressed into service to perform as a part of a company's brand, collapsing the distinction between the service offered and workers themselves. Service industry jobs like Rae's often require "emotional labor" wherein workers are expected to smile, be friendly, and even flirt with customers in addition to mixing drinks and taking orders.[1] Rae's use of her phone to disattend her surroundings, achieving a kind of absent presence, even if only while there weren't any customers to serve, suspended the rules of professional service work and allowed her to regain a sense of control over her work.[2]

While Rae relied on her phone to temporarily escape her workplace, Mike felt like he had to put his digital technologies away to achieve these same ends. Mike was a highly paid independent contractor in human resources for a national research company. He told me about how, on a recent family vacation while strolling on the beach, he accidentally dropped his work-issued Blackberry into the ocean. He uttered some words he wished he hadn't in front of his kids, chiding himself for bringing his phone to the beach in the first place. As he hurried to rescue the phone from the water, his mind raced with worry about missed emails and texts from colleagues about a project that was on deadline. He immediately turned the phone off and removed the battery, and as they continued their walk, Mike told me he began to feel "good" being technically disconnected from the office. Returning to his hotel room that evening, he popped the battery back into the phone, relieved to see its screen light up. He logged in to his email and found:

> There wasn't that much there . . . I caught up on a few missed emails, made a couple of comments on a document . . . my wife made some comments about "Not so bad after all, huh?" or something cute like that, but she was right. I ended up leaving the work phone in the room the next couple of days. . . . Why do they get to interrupt my vacation? It shouldn't be that way . . . I kept checking at night after [the kids] were asleep . . . I couldn't totally disconnect . . . They knew I was on vacation, but I had to look like I was

still doing my part . . . I mean, if nobody else does it, I can't do it, that would make me look bad.

Mike enjoyed the disconnection, leaving his phone in the hotel room daily to resist the pull of work during his vacation with his family. What had started as an unwelcome and accidental disconnection had prompted a deeper questioning of his need to be connected and desire for autonomy. But, as much as he might have wanted to completely shut off work during his vacation, he was worried about how this would look to his colleagues.

Mike's decision to leave his phone in his hotel room resisted the expectation that often faces high-wage professional gig workers to be devoted to their teams and projects on and off the clock. As a highly paid contractor, Mike was technically self-employed but relied on contracts with several long-term clients for steady income. Continuing these relationships for financial reasons exerted an important kind of control over Mike's work on his team, and the team's "culture" was another.[3] These two forms of control were in fact intertwined, as Mike's "fit" with the team was key to his continuation on the project. Despite his status as a contractor, Mike was still deeply affected by expectations, like being responsive to his coworkers outside of normal working hours, from his coworkers and manager as a signal of his personal commitment to his work. Leaving his phone behind was an act of resistance against these expectations of devotion to his team and project, albeit not without some compensatory work later on in the evenings.

In precarious work, a single outburst at a customer, missed call from a manager, or unanswered email can feel like an invitation to job loss and financial strain. In these labor markets, workers walk a tightrope between keeping their jobs and protecting their dignity against the daily demands of their clients and customers. For Mike and Rae, using and refusing their digital technologies were essential to resisting work and regaining some autonomy and control over their experiences of insecure work.

While management technologies have long been studied as a key part of worker resistance, the role Rae and Mike's phones played in their resistance to work went beyond the role these technologies played in their management. Labor process theorists have long pointed out that technologies of management and control can also facilitate worker resistance.[4] In classic studies of industrial labor, workers defended their dignity by breaking machines, worked around the imposed pacing of assembly lines, and subverted surveillance over their work.[5] In postindustrial economies focused on knowledge and service work, digital technologies are often seen as extending managerial control over workers outside the boundaries of the workplace.[6] However, the place of digital technologies in the labor process doesn't fully explain Mike and Rae's practices. As an independent contractor, Mike didn't have a traditional manager, and Rae's phone wasn't a part of how she was managed at work. Nevertheless, their phones still played a key role in the ways they resisted work.

Despite these similarities, digital technologies played different roles in workers' resistance. While Rae immerses herself in her phone to regain some control over her job, Mike has to actively plan to leave his phone in another location to do the same. What accounts for these different patterns? In Chapter 1, I explained how digital technologies were used in surprisingly similar ways by the high- and low-wage workers in this study. Workers all used their digital technologies as part of their "digital hustles," unpaid work done to find and maintain sources of income in precarious labor markets. This chapter explores a striking divergence in the different ways workers used and refused digital technologies—from phones and laptops to email and social media—to do so.

I found that different forms of labor control in high- and low-wage workplaces patterned the use of technologies in strategies to resist some of the dictates of their jobs and helped them claw back some of their autonomy at work. Low-wage workers largely faced forms of direct managerial control over their bodies and feelings, and they used their phones and social media to temporarily escape the constraints and expectations of their workplaces by immersing themselves in secret social spaces and suspending their professional performances. By contrast, the high-wage workers faced normative control, wherein management practices attempted to nurture shared beliefs and values, often related to devotion and attachment to the company or team, and a close personal identification with the job.[7] These workers resisted by refusing their digital technologies in order to limit responsiveness and affective commitment.

Through talking with workers about the place of their digital technologies in their hustles for precarious work, it became clear that their practices weren't only central for proving their worth as competitors in cut-throat markets but also in carving out zones of autonomy in their chaotic work lives. While studies of labor resistance often focus on collective action, I found that both high- and low-wage workers use their technologies as "weapons of the weak," enrolling them into individual strategies to resist managerial control.[8]

These types of strategies "require little or no coordination or planning; they often represent a form of individual self-help; and they typically avoid any direct symbolic confrontation with authority or with elite norms."[9] Workers' strategies were more aimed at altering their feelings about their work than they were about achieving more visible or collective goals.

Precarious work was supposed to enervate resistance. Sitting outside of legacy labor organizations and with conditions that keep workers fragmented and isolated, many have observed the bleak prospects of collective organizing among precarious workers.[10] However, in this chapter I'll show how contingent workers across the income spectrum resisted managerial control through strategies that were pegged to their digital technologies. By both using and refusing to use their digital technologies, workers struggled to create autonomy and dignity in insecure work through individualized, affective strategies where more organized and outright struggles may have felt impossible.

Absent Presence: Suspending Performances of Low-Wage Work

Like Rae, many low-wage gig workers felt that being disconnected from their phones at work wasn't only inconvenient for their ongoing digital hustles for work; it was also a threat to their strategies to regain agency and control within work contexts that are often hostile to it. These workers actively used their digital technologies in strategies to resist managerial control at work, immersing themselves in games or social media in order to suspend their performances and resist emotional labor. Their phones facilitated a kind of absent presence, where they could maintain the "impression of compliance" with bodily requirements to stay behind a counter or stock a shelf, or prohibitions against carrying their phones with them while they worked, while actually resisting this control by temporarily putting their attention and feelings elsewhere.[11] The ways low-wage workers used their phones and social media allowed them to be there while not *really* being there, being present in body, but resisting in practice.

After her shift at the upscale restaurant in Washington, DC, where she worked as a waitress and bartender, Kitty and I sat on an overstuffed couch in the cocktail lounge as she showed me exactly how she folds and ties the strings of her apron to conceal her phone in its front pocket, as well as the settings that keep it quiet but still alert her to calls or texts with subtle vibrations. At twenty-two, Kitty gave the impression of a wizened service professional beyond her years. As we settled into our interview, she told me that, along with a variety of other part-time food service gigs, she'd been working at this restaurant for four years, having started there immediately after graduating from high school. She was planning on a career in the restaurant industry and thought that college wouldn't serve her as well as understanding the industry from the ground up.

After chatting for a bit about her typical hours and the demanding clientele, I asked her if she used her phone while she was at work. She dug into her bag and pulled out her apron, explaining that service staff weren't allowed to have their phones with them as they worked but that management was more relaxed about enforcing the rule with the more experienced servers. In her years there, Kitty had learned "the rules behind the rules . . . you just don't pull out your phone when customers can see you, it's not a good look. It's just not professional." She pointed out that her manager expected her to have her phone with her both for emergencies and to look up recipes for the complicated cocktails customers asked for, though she'd usually "sneak to the back" for the latter because she worried it might ruin her performance as a knowledgeable upscale bartender. To keep her phone with her while still maintaining her professional performance, she developed a method for folding and tying her apron so that her phone was invisible but still accessible.[12]

As important as it was to do her job, Kitty's phone also played a central role in her resistance to managerial control over her emotions as a part of her job. Kitty explained that using Twitter on her phone to complain provided a "refuge" from the requirement to maintain a pleasant and deferential attitude at work:

Oh Twitter! There's "#hostessprobs" . . . sometimes I'll go there to seek refuge. . . . Occasionally I'll be like [mimics typing on phone] "this bitch server" or "this terrible customer blah blah" and sometimes people will reply, but even when they don't, I feel better because I got it out, as opposed to going in the kitchen and telling everyone in there.

Julia: Why not just go into the kitchen?

Sometimes it's just easier . . . sometimes it reminds me I'm a person outside the restaurant, you know? . . . It happened a couple days ago when there was this customer who came in and yelled at me for having music playing too loudly. . . . So I went on there and I tweeted, "This guest just came in and told me to turn down Earth, Wind, and Fire and you don't ever turn that down, you just talk louder!" . . . it was like three tweets long. I was just furious about it . . . so in between those three tweets, in that three minutes, all of a sudden I felt better. I didn't have to go yell at someone about it. . . . It's your own personal little soapbox that you can get on and you can bitch about stuff, if someone wants to reply, then great. . . . I can just step out-side Kitty the smiley bartender and be Kitty the funny Twitter bitch. . . . Sometimes it just feels better to put it on Twitter.

Kitty described the relief she felt after posting about a frustrating interaction on Twitter by using a hashtag to tap into a social space that exists outside the boundaries of her restaurant and where other service industry workers trade horror stories and share their gripes. Tweeting about her frustrating experiences was a reminder that she's a "person outside the restaurant" and allowed her to momentarily suspend her performance.

The creation of secret social spaces has long been a tactic for worker resist-ance, especially in feminized work.[13] Kitty's use of Twitter takes this tactic online, making use of the networked affordances of social media to gain public witness to her experiences alongside a community of other workers like her. For service workers like Kitty, posting to Twitter may have given her an immediate sense of relief, but it also gathers together a collective record of the otherwise individual experiences of these workers. The hashtag Kitty mentioned using includes tweets from others, such as:

"Some man just winked at me and I'm uncomfortable."
"Ending my night pissed off because some jackass thinks its ok to tip $5 on a $100 bill."
"What's the point in changing your availability if your managers schedule you whenever they want anyway?"

These tweets illustrate how service workers use social media to blow off steam and alleviate boredom and, as Kitty described, gain a sense of au-tonomy over the conditions of their work in a way that was often denied them in their high-supervision, low-autonomy positions. Some may point out that individual forms of resistance like this may be fleeting, or even more perniciously, allow service workers to accommodate themselves more fully to the demands of their jobs, normalizing exploitation and harassment.

Privacy Without Power

While workers like Kitty used social media to reach out to communities outside her restaurant, not all low-wage workers could dive headlong into their technologies to resist work. For workers negotiating marginalized identities in their workplaces, these "networked publics" presented new challenges to carving out zones of autonomy from work. While many interviewees, both high-and low-wage, felt pangs of uncertainty about their privacy while using social media, low-wage workers doing insecure work who were also marked as different by their race or sexuality felt a heightened sense of visibility because these marginalized identities followed them into online spaces and shaped their ability to use social media to resist work.

Veronica led me through a crowded restaurant in Washington, DC, to a corner booth and sat with her back to the wall, facing the crush of patrons grabbing a quick sandwich on their lunch break. Before divulging any particularly revealing information, she'd glance over my shoulder at the other tables, once musing out loud with a quiet laugh, "I hope I don't know any of those people!" As a retail salesperson in an electronics store nearby, Veronica used a Facebook group to trade shifts and information with her coworkers. As a queer woman who was active in her local gay community, she used social media to express political opinions, share local events, and gripe about work, with other members of this community.

Throughout our interview, Veronica talked fondly about her followers on Twitter, telling me about the complete strangers who follow and interact with her in that space. She recounted an instance a few years ago, when she attended a gay pride parade; while watching the festivities with friends, a tipsy stranger approached her in the street calling out her Twitter handle.[14] She felt a small thrill at being recognized in public, but this feeling of excitement over her "microcelebrity" status quickly transformed into stomach-churning dread.[15] Her manager had recently created a Facebook group to make it easier for her and her coworkers to trade shifts, and the event at the parade had called into question the strategies she was using to keep her personal life and work separate. She had erected what she thought was an impenetrable "firewall" between her personal and work social media use:

My Facebook is just for work pretty much, it's super generic. My Twitter is for everything else. Nobody at my job knows my Twitter.... I'm trying to get more hours, keep the ones I have, so I love using that Facebook group....

But I don't want something that I think is like nothing to end up being a big
deal for them. . . . You know, I'll talk about some ignorant stuff people say at
work, or complain about a customer, or post about who I think should run
for president or something. . . . On Facebook if you add where you work to
your profile, it automatically links you to that company's page so they can
see everything I write, so I did that there but don't write anything real on
Facebook. I don't associate with them [her employer] on Twitter. They don't
need to know everything about me.
Julia: What if someone just happens to find you?
They can't. On my Facebook you have be a friend of someone I'm friends
with to see anything. On Twitter, you have to know exactly what my handle
is, and it's not my actual name. . . . I like my personal life and my work life
separate . . . it can just cause problems. When people at work know your
business, it just gets too personal. I don't know these people, what they believe about being gay and whatever . . . and I want to still have a space to be
me, where I can just goof off without worrying about all that. So I just keep
it separate. Facebook for work, Twitter for fun.

Throughout the interview, Veronica described the ways she avoided "context
collapse" by managing this firewall; she obscured her usernames, avoided
posting photos of her face, and also conscientiously avoided sharing personal
information with coworkers while at work.[16] She also told me that the few
times it had been breached, such as at the gay pride parade, even though it
didn't involve her coworkers, made her "really nervous."

As a part-time employee, Veronica's hours changed from week to week,
sometimes getting cut or expanded without much explanation. She was always looking to pick up extra shifts from her coworkers, a process that was
previously coordinated through an opaque text message tree wherein her
coworkers would scramble around to figure out who could fill in, so partic-
ipating in the Facebook group was a boon to her bank account. She mentioned earlier that she undertook all these measures to avoid an incident in
which something unimportant or trivial that she posted would become "a big
deal" to her employer, risking a loss of hours or other opportunities. She also
mentioned her belief that getting "too personal" with coworkers can "cause
problems," but she was forced to adapt new strategies to keep this separation once her workplace started using Facebook to schedule shifts. As media
that connects different audiences and parts of our lives together, like social
media, become more embedded into everyday workplace practices, through

scheduling or exchanging information with coworkers, problems resulting from this "convergence culture" become obvious for marginalized workers.[17]

Low-wage workers like Veronica deployed these tactics as "weapons of the weak," but, as Lacey, a Black administrative assistant living in Washington, DC, explained, concerns about privacy may also shut workers out of their workplaces' secret social spaces. When I asked Lacey if she ever used social media for work, she let out an exasperated sigh. I had clearly touched a nerve. She explained that the other women in her office, mostly other administrative staff like her, were a close clique in the office. When she first started working there, one of them approached her in the break room and asked if she was on Facebook, explaining that they had a group "just for fun" to order lunch together, gossip about their bosses, and share information. Lacey explained that she didn't really use social media and ended up excluded from the social backstage of her office. However, she had absolutely no regrets about being left out. She told me that, while years ago she had a profile on BlackPlanet, she doesn't currently use any social media:[18]

I'll just complain to my real friends face to face, I don't need any social media to talk stuff about people I work with. I think it's stupid. I don't want everybody to know me. I don't want everybody to have access to me. . . . I feel like you could use that against me and then it could come back in some way to bite me in the ass. . . . For me to be minding my own business, just working hard to be where I'm at, working for every last little thing I've got, and then one day some comment I made is posted up somewhere on the news or on some site and I'm like "How the hell did I get here?" I don't know why the hell it bothers me so much, but when I hear stories about the Ray Rice thing, or the Beyoncé thing in the elevator. They were just doing their own thing, they worked hard and they were successful and then somebody sold that [smartphone video] footage to somebody else. You want money, so you want to sell your soul and sell mine for a million dollars. . . . I just don't want to give anyone access to me. I don't want to be on somebody's Instagram. I worked hard to get here, and I like to be able to control the access that someone has to me.

Lacey declined to participate in the secret social space created by others in her office over concerns that she couldn't really trust that any information she shared there would stay there. Lacey's suspicion of social networking as giving others "access" to her is couched in several high-profile cases of celebrity video leaks. To Lacey, these scandals are evidence of the vulnerability of these hard-working celebrities to the whims and greed of those around them. As she reiterated several times during her interview, Lacey has worked hard to get and keep her job, and as a single mom, she's careful not to take any risks that might jeopardize her hard-won security. She preferred to keep her resistance to work off the Internet and share her grievances in person.

Having their personal privacy respected isn't a default setting that Lacey and Veronica can assume they'll be granted. Intrusion and exploitation of personal information aren't surprises; they're regular features of their technological experiences, ones they must proactively guard against when it comes to resisting managerial control and carving out spaces of autonomy for themselves.

Privileged Refusal: Resisting Normative Control in High-Wage Work

Throughout my interviews with higher-wage precarious workers like Mike, it often felt like we spent more time talking about refusing to use their digital technologies than about using them. As independent workers, they keenly felt the lack of what they thought of as "normal" markers in time and space that put limits on their work days. From self-help and meditation to rigid scheduling and physically separating themselves from their technologies, they often talked about taking (or *wanting* to take) "breaks" from their digital technologies to create separation between their work and personal lives.

Given that they were in charge of their own schedules, this might not seem like too much of a challenge, but workers still frequently talked about feeling unable to put their phones down or close their laptops. High-wage workers wanted to be less connected to their work through their digital technologies, but in their efforts to limit their connectivity, they ran into a dilemma: connectivity was their primary means of signaling their devotion to the project or client, and efforts to limit it felt impossible. Instead of viewing these frustrations as the dark side of self-employment, I began to understand these acts of resistance as efforts to refuse work under conditions of normative control. Unlike the low-wage workers, who dove into their phones to carve out spaces of autonomy for themselves, high-wage workers more often refused to use their various technologies to do the same.

When I asked Jeff, a writer who juggled a nearly full-time corporate client and several freelance writing and teaching gigs, about whether he was "on call" to his many jobs, he responded:

I don't know how to answer that because I haven't yet figured out when I'm off! I like what I do. If anybody calls me, honestly, from any of my jobs, I'm usually pretty responsive. If someone called me on a Friday night and said I really need you to work on this, I'd say, "Okay, no problem, when do you need that by?" because I like it, and I like them, too . . . and [laughs nervously] I'd also like them to keep hiring me . . . it's not the only reason, but it's always in the back of my mind [sighs].

Jeff's hustle, which included tending to his multiple phones throughout meals and overnight, was a result of his enjoyment of his job, the people he worked with, and also the insecurity of his income from contract to contract. Jeff both wanted and needed to immediately respond to the many requests from his clients and students. While the concept of "resistance" at work is usually understood as resistance to managerial control—that is, to a source external to the workers themselves—in the case of these independent and insecure workers, figuring out what exactly they're resisting is a bit more complicated. It appeared that they weren't resisting control exerted by others but the pressure they put on themselves to constantly perform their devotion to their clients.

Like Jeff, the high-wage workers articulated the desire and need to make sure their clients knew they were committed to their projects and devoted to their success. Examining technology companies in the 1990s, Gideon Kunda observed what he called "normative control," where companies fostered organizational cultures that encouraged their employees to deeply identify with their jobs and devote themselves to achieving organizational goals.[19] However, as freelancers, many of these workers didn't have an organization to identify themselves with and instead thought of themselves as a "company of one," applying the same forms of normative control to themselves.[20] Foucault describes this "enterprising self" as a model of selfhood that emerged alongside neoliberal capitalism and implores all workers, not just those working independently, to become "entrepreneurs of themselves" by cultivating their own human capital and taking responsibility for their own economic livelihoods.[21] Rather than being controlled by a manager, high-wage independent workers must manage themselves, governing themselves according to the aims of their clients and the markets for their labor. This self-government was reinforced by their own self-expectations, and also those of their teams and colleagues.

Like Mike earlier, Jeff explained that he paid close attention to the ways his colleagues at his corporate client worked, and saw that they signaled their commitment to their work by logging long hours from home at night. He recounted noticing a "campaign" from the Human Resources staff about employees' "work-life balance" around the December-January holidays. In a series of emails, employees were entreated to not answer emails late at night or on weekends, but Jeff still felt pressure to log in after typical work hours to demonstrate his commitment to this client:

> You can see when people are online because the green light next to their name lights up. You can tell that people are on all the time, but there's no stated expectations for that. In fact, everyone says, "Don't be on all the time," but then they are on all the time. So I was like, if everyone else is doing it, maybe I should be doing it, too. . . . So there's no expectation, nobody would ever say it . . . it's just all in the background.

Earlier, Jeff described a prisoner's dilemma of connectivity. Despite his own desires to log off and limit the amount of time he spent online, he felt his own practices were constrained by his colleagues, whose practices reinforced connectivity as an expression of their devotion. Among highly motivated knowledge workers like Jeff, flexibility in the time and place of work tends to lead to both greater worker control and overwork.[22] In their study of high-wage tech contracting, Barley and Kunda explain that pressures to maintain a good reputation gave clients considerable control over contractors to log long and unpaid hours in lock-step with their salaried coworkers.[23] Many attribute the extension of work into the home to the expansion of personal digital technologies that allow workers to bring their work home and work from anywhere. However, in their struggles to resist work, I found it wasn't as much about the work their digital technologies allowed them to do, but the meaning of "logging in" that signaled devotion and commitment to work.

Jeff told me that he'd leave his multiple phones by his bed at night, with audible email notifications on, and often check them while bleary-eyed on trips to the bathroom in the middle of the night to make sure there was nothing new. He was rueful about these practices, smiling through several deep sighs as he told me:

Checking my phone like that is so draining. I know it's bad to do, and I wake up tired in the morning and feel bad about it, even though I'm working and it's not like I'm wasting time or something. I would like to have more rules. [Sigh] I would like to have more rules. . . . I was into it for a while, I had strict, like I set an alarm for 9:30 p.m. and was supposed to put everything away . . . laptop, phone in the nightstand. . . . I was listening to this self-help type talk about unplugging and meditation and it made me realize that people are too easy on themselves. . . . You've got to grab hold of it because you're in charge of it and, in the same way that if you don't go to the gym and you get flabby thighs, there's nobody really to blame but yourself. . . . It's just self-discipline. . . . It lasted for about a week. . . . I think about this stuff a lot, [sigh] I think about it a lot. It's just hard to keep it up.

Jeff felt "drained" and "bad" about checking his emails at night, and he had tried to limit his connectivity by imposing strict rules through scheduling the time spent with his digital technologies. Jeff's alarms point to a nostalgia for an era of more stable working hours, where there were clear distinctions between working and nonworking hours. These attempts to tame time are also attempts to cope with precarity. As Melissa Gregg observes, white-collar workers' obsession with time management can be seen as "material and psychological support for jobs and careers that are felt to be unstable, improvised, and forever running at a frantic pace."[24] This tactic was unsuccessful, but he wanted to keep trying to cultivate the kind of self-discipline that would keep him from engaging in this "bad" habit. He turned to self-help, which reinforced this sense of individual responsibility for "grab[ing] hold" of his technology practices. Jeff felt that he, and he alone, was to blame for his overconnected working style and that better self-discipline was the solution.

Like Mike's strategy of leaving his phone in his hotel room, many of the high-wage workers made themselves physically or digitally inaccessible to cope with the downsides of constant connectivity. For Eleanor, an IT security consultant, vacations with friends required advance planning for her to disconnect from work in the way she desired:

I want to have a low-tech vacation, which has been my goal for a while now. . . . It's good to disconnect from the always on, pounding feeling that the email gives me at times. I wanted to cut off that responsibility. . . . I know if I bring my computer [on vacation] I'll be tempted to just check in once or twice in a way that's entirely unnecessary, and I'm always watching all of my friends checking in unnecessarily, too. They're always less than arm's length from work stuff that way.

Eleanor explained how she planned in advance to leave her laptop at home by buying an e-reader, which had limited Internet access, and downloading some books and movies so she could "cut off" the "responsibility" she felt to answer emails while on vacation with her friends. By purchasing additional digital devices, high-wage workers engage in "boundary work" designed to separate their time into working and nonworking time.[25]

Simply disconnecting from one's digital technologies is a privilege that is not available to all. As I explained in the last chapter, inaccessibility wasn't possible or desirable for many low-wage workers because of their precarious grip on consistent income. But it's not only economics that keeps low-wage workers connected and allows high-wage workers to take breaks; it's also their status at work. Restricting people's access in this way is a privilege afforded those with high job status.[26] The ability to ignore phone calls, to leave emails unanswered, and to "disattend" others' requests depends on status and access to resources, both cultural and material, and indeed may even serve to reinforce that status.[27]

Tethered Carework: The Limits of Refusal

Many of the high-wage workers I interviewed told me about expectations of 24/7 connectivity to colleagues and clients; however, for some, these expectations weren't only emanating from work but also came from home and family. While many of the interviewees described earlier used refusal as a tactic to separate work and their personal lives in ways that would prevent work from colonizing their lives, for high-wage mothers, connectivity often threatened to spill over in the opposite direction, allowing family life to flow into the time and space of work. High-wage workers' ability to disconnect from their digital technologies was shaped not only by their capital but also by gendered norms surrounding carework and communication.

Mothers face intense personal and individual pressure related to the many different aspects of their social role as parents.[28] Among both the high- and low-wage workers, I found that these expectations took a form I call "tethered carework" around digital technologies. People practiced tethered carework when they insisted on the importance of simply "being available" or "being connected" to their loved ones, usually by having their phones nearby, switched on, and monitoring incoming messages and calls. This form of digital kinkeeping, much like its many offline forms, is gendered labor usually taken on by women.[29] Tethered carework required them to be available and connected just in case they were needed by someone in their close relational networks.

Tethered carework made it difficult for high-wage mothers to keep their connectivity from spilling over the boundary they wished to construct between their work and family lives. Over coffee, Sirrom, a forty-two-year-old mom of an eight-year-old son living in Washington, DC, told me a story about an uncomfortable moment at her previous job where she had worked with another freelancer friend to provide human resources seminars for a large data analytics company. This company was her largest client, and she and her colleague had an understanding that Sirrom would only teach while her son was at school. However, when her colleague got sick, Sirrom stepped in to lead a seminar that overlapped with the time her son got off the bus after school. While normally Sirrom would be there to meet him, that day, her eight-year-old would need to let himself into the house and stay by himself for an hour before Sirrom got home. Back in the classroom, Sirrom was keeping an eye on her phone, both to keep herself on schedule with the curriculum and to make sure everything was alright at home. Suddenly, in front of a room full of students, she saw her phone light up with an incoming call from her son, and she had to think on her feet to quickly handle the situation:

I saw it and I'm like, "Oh, man. What if something happened at school? What if he's locked out?" I had us take a break earlier than what we should have, but it was like "I got to take this." . . . I stood outside, some people came outside . . . and they saw me on the phone. I was worried they might think, like "Oh yeah, take a break so you can make your phone call." Then, the vice president walked by . . . and in my mind I'm like "That is not a good thing." But nothing was ever said; it was just me making myself crazy. I felt embarrassed; I felt a little incompetent.

Julia: What about it made you feel incompetent?

I'm using business time for personal matters. . . . Does he [the VP] think I'm less committed to the job? Does he think I'm not as good of a trainer? Does he think that I have personal things that are going on that are impeding my work? But I overthink everything, so that's my fault. . . . I'm just shaking my head at myself, like "Really? Don't you know better?" I should've run outside to take that call, but I didn't even think because I was worried that something had happened; it short-circuited my thinking in that moment.

Workplaces are emotionally fraught contexts, especially for women with care responsibilities.[30] They navigate competing cultural messages that reinforce the necessity of their devotion to work, on the one hand, and devotion to family, on the other.[31] For white-collar mothers coping with precarious work, connectivity didn't only perform devotion to work but also to their families.

Sirrom experienced tensions between gendered expectations about being available to her son and workplace expectations about what she owed her boss and coworkers. These tensions are always there, but ubiquitous connectivity facilitated by phones may allow them to surface more poignantly, more urgently, and more often. Managing the phone, and the tugs of these competing expectations, is a lot of work; but it also takes an emotional toll.[32] Sirrom's strategy for dealing with the competing demands of work and family was to put the class on a break to take the call from her son, but once she deployed this strategy, she blamed herself for both the choice and the potential fallout from it.[33] In doing so, she transformed problems emerging from dueling cultural expectations and context-based constraints into individual problems that she alone was responsible for solving.

While men also sometimes faced expectations around tethered carework, they felt much less tension in asserting strong boundaries to accommodate their preferred ways of working. John was a forty-eight-year-old independent financial advisor who ran a successful firm out of a small office space in Mainville, New York. John was an outgoing and gregarious man, joking and chuckling through his interview, especially when he talked about how he dealt with the "constant chatter at me" from his female relatives:

> like when my sister will send me a text. If I don't answer it within a couple of minutes, I get the capital letters "hello" text with a million question marks. That's when I call her and say, "I'm fucking busy, I can't answer you right now!"
>
> Julia: Do you ever turn it off?
>
> Well, I'm criticized for that. I leave it off all the time, so they criticize me for it. They're always saying, "What if something happened? What if we need to call you?" I'm not attached to it. . . . I don't like small talk while I'm at work, so I don't use it just to call and chitchat.

John's confrontation with his sister and the ease with which he explains his practice of turning his phone off, against the wishes of his sister and other family members, contrast strongly with Sirrom's self-doubt. John has children about the same age as Sirrom's son, but he felt unconflicted about their ability to get in touch with him during the work day, characterizing these interruptions to his work as "small talk" and "chitchat" that distract him from focusing on his clients.

While it's important to recognize that simply disconnecting from work to enjoy their free time is a privilege afforded to only some workers because of their social status, it's also necessary to see that this privilege is also a gendered freedom. The ways high-wage hustlers were able to control their connectivity and stop it from spilling over the boundary they had constructed between their work and personal lives were inscribed by gendered expectations of communication and care in families. High-wage independent workers had many advantages, but some were still constrained by social inequalities that shaped their hustles.

Conclusion

Over the past fifty years, the US economy has shifted from one that specialized in the mass production of consumer goods to a polarized one focused on highly paid knowledge and low-paid service work. This same period has also seen the erosion of stable employment conditions and the spread of economic insecurity for many Americans. Digital technologies have taken center stage in our understandings of these shifts. From the ubiquitous use of mobile devices that extend managerial control far outside the walls of the office, to surveillance and automation technologies that tighten the grip of managerial control on in-person work, digital technologies are central figures in creating alienating conditions of work. However, beyond their role in making work harder for workers, digital technologies also play a key role in the ways workers resist work and carve out autonomy for themselves.

Despite the fact that technologies weren't a major way they were managed, I found that high- and low-wage contingent workers used and refused their devices to resist work. While low-wage workers dove into their phones, high-wage workers refused their phones and other tech to create autonomy. This difference in strategies is explained by differences in labor control. Low-wage workers faced direct control over their bodies, behaviors, and affect, and used their technologies to suspend their professional performances and create secret social spaces. High-wage workers, while they often lacked traditional relationships with managers, still faced forms of normative control over their work that emphasized their devotion and commitment to clients and teams. To resist work, they took breaks and refused to use heir digital technologies to limit their responsiveness and availability.

Networked digital technologies have sparked hope as new spaces for labor activism and worker voice. From on-demand gig workers using Facebook groups to organize strikes, to the use of social media to draw attention to the labor struggles of teachers, Amazon warehouse workers, and Uber drivers, workers have used social media in their fights for their rights.[34] Workers' use of social media and mobile phones in both institutionalized labor politics and visible public debates about their rights have shown that these technologies aren't only tools for managerial control but important tools in workers' fights for dignified and safer work. However, this celebratory narrative overlooks the role of digital technologies in quieter, more individual, and more pervasive forms of resistance against work.

Beyond legislation, lawsuits, and viral grassroots activism, worker resistance also includes these more individual and covert actions. These forms of resistance are a "struggle to control the concepts and symbols" of low-wage service work in the context of precarious employment, where more collective and visible struggles may feel more difficult.[35] For workers like Kitty and Rae, using their digital technologies to create secret social spaces, gain public witness to their experiences, or to just avoid boredom, despite prohibitions was a small-scale contestation of the managerial control over their bodies, feelings, and connectivity while at work. This use of their phones may provide a sense of empowerment over the rules of professional performance, while at the same time, leaving those rules unchallenged.

While digital technologies enabled some workers to claw back a space for autonomy at work, these strategies may also reinforce existing inequalities among workers. As this chapter describes, using and refusing digital technologies in strategies to resist work wasn't an option available to all workers. Taking extended breaks from email and social media or using social media to connect with other workers or even as a temporary distraction from boredom were privileges conditioned by both racial and gendered privileges surrounding surveillance and carework.

As celebrities like Chrissy Teigen and politicians like the mayor of Barcelona announce "breaks" from social media (ironically, using social media) and take extended "sabbaticals" from work and email, they offer a critique of the expectations of connectedness required for their digital hustles.[36] However, this critique only serves to reinforce their status as busy and sought-after professionals, and the falsehood that individual will is responsible for their successful management of floods of connectivity that comes with intense workloads. Many scholars have conceptualized how flexible and yet still "always-on" work cultures constrain high-wage workers but comparing these workers to their lower-wage counterparts allows a wider view that throws their relative privilege into sharp relief.[37] These strategies aren't available to all, as the performance of tethered carework remains a gendered expectation of mothering and kin-keeping in our digitally saturated lives.

In addition to gender, race shaped the conditions of possibility for workers to use their digital technologies to resist work. As Simone Brown explains, while impositions on our privacy facilitated by digital technologies may feel new to many, "surveillance is nothing new to black folks."[38] As social media and other connective digital technologies become more widely used and deeply integrated into workplaces, both in efforts to extend managerial control and in workers' resistance, it's increasingly important to take an intersectional view of power to acknowledge and mark the ways that workers' abilities to use their digital technologies to resist managerial control are shaped by the intersecting oppressions of homophobia, racism, and sexism.[39]

While many have celebrated the possibilities that new digital tools offer for collective organizing and new forms of worker voice, access to digital technologies alone isn't enough to produce these outcomes. Even if they could, these tools aren't available to all workers in the same ways. In the context of widespread precarious working conditions, it's essential to recognize the potentials and pitfalls of forms of resistance beyond collective organizing, if only to recognize the myriad ways that vulnerable workers can and do exert their power.

Conclusion

Beyond Inclusion

Days before the 2021 Super Bowl, Squarespace, a company that sells website building and hosting, released an ad featuring Dolly Parton singing a remake of her classic work anthem, revamped for the gig economy, titled "5 to 9." With an estimated one hundred million viewers, Super Bowl ads are some of the most expensive and visible real estate on television today, and companies paid over five million dollars for a thirty-second spot.[1] Parton's original song was the title track to a movie that spawned a national awakening about issues of sexual harassment, pay disparities, and bad conditions in feminized clerical work in the 1980s. In this playful reboot, Parton sings an ode to pursuing one's passions outside of traditional employment. At a time when digital technologies have become infrastructural to coping with precarious work, it's hard to imagine an ad that more poignantly captures the cultural contradictions at the heart of the digital economy.

The commercial opens on the drab interior of an office: workers struggle to stay awake in a sea of cubicles, clicking away at desktop computers and slamming down chunky black plastic phones into their handsets. The office clock ticks over to 5:00 p.m. and a sleek sliver laptop is suddenly opened, fingers eagerly tapping out edits to a colorful personal website. Parton sings, "Working 5–9, you've got passion and a vision, cuz it's hustlin' time, whole new way to make a living. Gonna change your life, do something that gives it meanin' with a website that's worthy of your dreamin'." Each nondescript beige cubicle is transformed into a colorful workshop for a baker, an artist, a woodworker, as formerly listless workers dance and use sleek tablets, phones, and laptops to seamlessly bake a cake, sand a chair, and edit and update their websites. The viewer follows one woman, dancing in bright yoga pants among her newly invigorated colleagues; she enters an elevator and the doors close as she glances at a website for her dance fitness business on her phone, smiling broadly as she leaves the drab office behind as the tag line "Make 5–9 full time" appears.

Left to Our Own Devices. Julia Ticona, Oxford University Press. © Oxford University Press 2022.
DOI: 10.1093/oso/9780190601288.003.0006

This ad and Parton's song celebrate the contradictions of the relationship between precarious work and digital technologies. While most of the contingent workers interviewed for this book probably wouldn't say that their smartphones, laptops, and Internet connections were helping them live out their dreams, these technologies were a part of the ways they constructed their identities and found dignity in their day-to-day experiences. The ability to win new clients and hop from gig to gig, seamlessly communicating with customers and adapting to constantly changing circumstances, helped create some certainty about their abilities in very uncertain economic circumstances.

At the same time, the use of these technologies, along with the Internet, to engage in continuous work (from 9 to 5, and 5 to 9) is also a response to stagnating wages, a withered social safety net, and to the shifting of risks of employment onto workers. These conditions have allowed for the exploitation of workers' connectivity, as our institutions simultaneously require, ignore, and punish the use of digital technologies by marginalized workers. The role of digital technologies in our current system of contingent and precarious work is both exploitative and empowering because these technologies are being used to solve problems at the heart of this system. These are the ways we work when we're left to our own devices. Examining the role of digital technologies in precarious and contingent work exposes contradictions that must be accurately seen in order to be understood, and in order to reckon with the complex politics of our current moment.

* * *

This book is not about the future of work. Over the past several years, speculations and predictions about the evolving role of digital technologies, algorithms, artificial intelligence, and platforms in the world of work have taken center stage. Boosters welcome these changes and look forward to disruptions that will enhance productivity, democratize access to opportunity, and create new, more flexible, relationships to work.[2] At the same time, critics rightly warn about new forms of surveillance, discrimination, and the tightening of employer controls over workers' daily lives that these same changes portend.[3] But, as I sit writing these words, our world is at what might be the beginning of an extensive period of disruptions to work, family, and civic life caused by the COVID-19 global pandemic, and suddenly the future of increasingly digital, dispersed work is, for many of us, our present reality.

Other workers have either had their access to income abruptly cut off or are making painstaking tradeoffs between their health and economic survival.

This crisis has shown, maybe more clearly than previous disruptions, the necessity of connectivity to the Internet for the execution of daily life and the increasingly untenable assumption that this should be left up to individuals to provide for themselves. We are all affected by these digital inequalities. It's not only a problem for those who struggle to maintain connections; it's a problem for those who rarely have to think about it as well.

The current political economy of connectivity is set up to allow some of us to forget and ignore others' struggles to access the Internet, phone, and data services. It's important that we reckon with everyday forms of digital privilege that allow some of our incomes, schooling, and groceries to be delivered smoothly. Ignoring this privilege is one of the moral hazards of living in precarious economic times. When we don't recognize the unearned assets that make using digital technologies seamless for some, it perpetuates the idea that these technologies democratize access to economic mobility for all, and it contorts our ability to see how digital technologies can exacerbate inequality. When we can't (or won't) see how our own privileges allow some to thrive and leave others to fail, we continue to believe the myth that anyone with a smartphone who really wants to can use it to pick themselves up by their own bootstraps. When we can't look beyond divides to see how digital inclusion can systematically advantage some and marginalize others, we endorse the exploitative logic that exists now and neglect the ways that our fates are all tied together.

* * *

Chapter 1 described a set of practices, the digital hustle, that is shared among low- and high-wage independent and contingent workers across many different types of work. These practices include maintaining vigilant watch over email inboxes, text messages, and battery life in order to secure and coordinate paid gigs. But it would be a mistake to assume that the frenetic practices of the digital hustle are driven only by economic needs. Right alongside the need for income is a deeper, perhaps even more pressing need for dignity and reassurance that many gig workers lack as they face the labor market mostly alone. Beyond income, I found that the digital hustle is also central to the construction of workers' identities as worthy competitors in cutthroat markets. The practices of the digital hustle produced both income and pride for high- and low-wage workers, leading me to identify it as a craft practice of the gig economy.

Although the hustle is shared, it doesn't have the same consequences for workers at either end of an increasingly polarized labor market. Chapters 2 and 3 point out the vastly different circumstances that confront low- and high- wage workers as they construct and execute their hustles. Chapter 2 argues that to understand how digital technologies can still amplify inequalities in gig labor markets, even as more people get online, we need to go beyond looking at exclusion and examine the terms on which people are included into the digital economy. This chapter points out that, far from being excluded, low-wage workers are increasingly expected to be connected. However, the terms on which they're included reinforce and amplify existing social inequalities. In their efforts to take advantage of low-cost phone plans and free online services, low-wage workers confront companies that make connectivity appear affordable, but actually extract high interest rates, and share their personal information with advertisers. This fragile connectivity is shored up by workers' invisible labor, like negotiating with salespeople for extra time, foraging for free Wi-Fi, and juggling multiple devices, that compensates for routine disruptions.

While low-wage gig workers struggled to maintain connectivity, high-wage workers seemed to have too much of it. In Chapter 3, I argued that, instead of understanding white-collar gig workers as having superior skills, we need to examine the ways they benefit from digital privilege produced by the social field of gig work, which allows them to execute successful hustles in ways that look like individual achievements. Their connectivity is both expected and rewarded by their clients and workplaces and facilitated by their economic and social capital.

Their digital technologies were essential tools to get and keep their work, and they were also central to how both high- and low-wage workers resisted work. Amid heady optimism around the promise of digital tools for collective labor organizing, Chapter 4 points out that workers' use of digital technologies to resist work wasn't aimed at the role these technologies played in the labor process, or reaching out to build collective solidarity with others, but centered on a struggle to control their feelings about their work. These strategies were useful for some, allowing them to claw back some autonomy, but they weren't available to all. Taking extended "breaks" from social media and email, or even using them to reach out in the first place, is a racialized and gendered privilege, conditionally extended to those unmarked by expectations around caregiving and histories of unequal surveillance.

Talking with workers from across industries and at either end of a polarized labor market has shown me that our increasingly future-oriented discourse about work and technology is obscuring crucial realities about work in the current moment. This is a book about the conditions currently facing gig workers and others with nontraditional relationships to work and the ways they navigate relationships using the digital tools they have at hand. I don't know what the future holds, but a present that's this complex will likely lead to a future that is equally if not more complex. However, this book also provides an example of how we can better understand these complexities moving forward. Looking at the similarities and differences in the practices of digital hustling and the conditions confronting high- and low-wage gig workers illustrates an approach to seeing inequalities in complex sociotechnical systems. This case study provides a few lessons about how we can better understand these issues moving forward.

Precarious Work beyond Platforms

This book has focused on gig workers, including full-time freelancers with their own small businesses, those working in informal under-the-table arrangements, and those with temporary employment or who are moonlighting by combining traditional full-time and contingent work. The diversity of work arrangements I encountered, even in this single study, defies easy categorization. This reflects the heterogeneity in the landscape of independent and contingent work in the United States today.

Oftentimes, in our discussions about gig work, the definition narrows to workers who find work through online platforms or apps like Uber, Instacart, or Amazon Mechanical Turk. Given the obvious role of digital technologies, Internet connectivity, algorithms, and big data surveillance in these types of work, the experiences of these workers are important to understanding potentially larger shifts in our ever-more digitally entangled working lives.[4] But, as this book has endeavored to illustrate, these workers are simply the tip of the iceberg. Online platforms have opened the door to gig work for many, but these companies are just one chapter in a much longer story about how symbolic boundaries are drawn between formal and informal, skilled and unskilled, visible and invisible kinds of work. These lines have been drawn in different places over time and have rewarded workers with more status, better pay, legal protections, and entitlements while punishing others with wage penalties, unsafe workplaces, cultural invisibility, and even moral condemnation for presumed lack of individual skill or self-discipline.[5]

As I illustrated in Chapter 2, workers on the "wrong" side of these boundaries are often assumed to be unskilled, disconnected from digital technologies, or using them in ways that distract them from work. This chapter illustrated that digital technologies are no less essential for workers coordinating gigs as movers, landscapers, and eldercare workers than they are for white-collar workers, laying bare that, as Virginia Eubanks observes, "what counts as 'real' interaction with IT is defined in profoundly class-, race-, and nation-specific ways." These workers are no less digital than those who rely on online platforms to find work, but our definition of them has profound consequences for who is included and excluded from conversations about the role of digital technologies and work.

For example, online platform companies, rightfully, draw our ire when they steal tips or force workers into unsafe situations, but so should employers of low-wage workers who expect them to use an app or text messages to receive their weekly schedule or a company website to access benefits or other resources but don't subsidize their connectivity. These apps and systems sometimes make life easier for vulnerable workers—being able to swap shifts through an app is much easier than calling and texting your coworkers individually—but maintaining expensive devices, cell phone service, and data plans becomes a new kind of tax on low-wage workers' already stagnating wages.

Narrowing our scope for understanding the ways digital technologies are affecting low-wage work to on-demand platforms also forecloses critical scrutiny on the disciplinary uses of a wide array of technological systems. From customer surveillance systems in retail stores that also serve to control workers to automated job application platforms, increasingly individualized surveillance technologies are becoming more commonplace in low-wage work and are deeply tied to workers' cell phones and broader online presence.[6] It's crucial that we widen our perspectives on the growing integration of low-wage work with digital technologies so that we can identify otherwise hidden connections between technologies that both facilitate work and seek to enclose and discipline workers.

Unequal Inclusion: Institutions and Digital Inequality's Middle Range

In order to see these countervailing forces, it's also important to be attentive to digital inequality's middle range—or the diverse set of institutions, organizations, and contexts that mediate individual interactions with sociotechnical systems. These social formations make up what's often called the "meso" or "middle" level of social organization—occurring somewhere in between the "micro" level of individual interactions and "macro" level formations like nations. In Chapters 3 and 4, I illustrated the ways that the different contexts of low- and high-wage work shaped whether workers were rewarded or punished for their digital hustles. I also pointed to the role of public and private institutions like libraries, fast-food franchises, and coffee shops in both facilitating and disciplining the connectivity of different types of workers. This book joins a growing chorus of scholars arguing that digital inequalities are constructed through our relationships with many different kinds of institutions, from political and community organizations, to workplaces, schools, and families.[7]

For nearly two decades, the ways we've examined and explained digital inequalities has made it more difficult to see these middle-range actors. Digital inequalities are often understood as resulting from exclusion from access or meaningful use of digital technologies. Scholars have examined problems that arise from some people having access to the technologies, services, and skills they need to fully participate in society while others do not. Scholars ask questions about the ways socially patterned differences between individuals, such as in income, technology skills, or education, result in the exclusion of some groups from access to and use of digital technologies.[8] However, digital inequalities are not only problems of exclusion.

More broadly around issues of digital technology, "inclusion" has become a buzzword for efforts at broadening access and ensuring diversity in designers, engineers, and users.[9] But increasingly, there's another kind of inclusion that has also drawn our attention; we are being included in sociotechnical systems of work and social life that use vast amounts of data to surveil, track, and categorize us whether we like it or not. The discourse of inclusion as broadening access often obscures the fact of inclusion as surveillance and control of marginalized populations.[10] For those who care about digital inequalities, there are urgent questions about the conditions under which groups are included in various kinds of mandatory, potentially discriminatory or harmful sociotechnical systems, not only the ways groups are excluded from beneficial uses of digital technologies.[11] Digital inequalities scholarship, and those seeking to fight these inequalities, must be able to see the crucial role that institutions and organizations play in producing the vulnerabilities that stratify society.

The colloquialism I borrowed for this book's title, being "left to our own devices," which signifies being left alone, without help from or, conversely, being controlled by others, is not only a good explanation of our intimate relationships with our digital technologies but also an apt description of the wider pattern of the waning role that many institutions, from the state to the workplace, are playing in the social reproduction of daily life. This is the story labor scholars have told about the path toward precarious labor paved since the 1970s as social contracts between employers and employees have weakened and eroded previously taken-for-granted entitlements to health care, retirement plans, and workplace safety.[12] However, the rise of precarious labor isn't only about the shrinking sphere of responsibility that employers have defined for themselves; it's also intertwined with the expansion in the role of other institutions, from new types of companies to the criminal justice system.

New and transformed institutions are stepping into this breach left by traditional employers, carving out territories from what's been left behind to create new regimes of digital inclusion. There are both liberatory and predatory sides to this inclusion. Predatory forms of digital inclusion exploit our reliance on digital technologies and data-intensive services to profit in ways that depend on and exacerbate other kinds of social inequalities. This includes, as I uncovered in Chapter 3, seemingly affordable but actually exploitative financing for smartphones, enrollment into surveillance systems without consent, and the use of personal data to target us with predatory products and services. The idea of predatory inclusion prompts us to ask questions, not about the barriers to closing the digital divide, but about the interests that are served by keeping it open.[13] By contrast, liberatory inclusion leverages our constant connectivity toward overcoming and reducing existing social inequalities through, for example, new kinds of labor organizing that bring workers together, or mobilizing networks of grassroots mutual aid through social media.[14]

Paying attention to digital inequality's middle range makes it possible to see how digital technologies can build both inclusivity and exclusivity at the same time. As connectivity to Internet, data, and messaging becomes essential for workers up and down the income ladder, it is becoming clearer that we're seeing patterns of segregated inclusion, where new social groups are granted access to previously elite resources, and simultaneously, new kinds of exclusivity are born.[15] For example, when low-cost phone plans offer "unlimited data" but throttle their users' speeds during peak periods to prioritize providing faster Internet to their wealthier customers, this is segregated inclusion. Labor market demands to be connected have brought formerly separated groups—like highly educated freelancers and low-wage gig workers—together on the Internet, but in ways that allow elites to maintain their power and distinction.

Eliminating economic barriers to high-quality connectivity is important, but it's also worth pausing to ask about what kind of Internet we are fighting to include people in. The twin faces of predatory and liberatory inclusion draw our attention to the ways that full inclusion into a digital economy that favors the interests of technology companies, data brokers, and Internet service providers over the interests of workers while enabling extensive surveillance and profiling of vulnerable populations is exactly what creates massive profits for them. The possibilities of liberatory inclusion make the struggle to connect everyone important, but the perils of predatory inclusion make efforts to secure our privacy and fight discriminatory data profiling necessary partners in ensuring equity online.

Studying Digital Privilege

After their interviews, some people asked about the research project, who else I was interviewing, and what I was finding. After confessing he was a sociology major in college, Leonard, a young social media marketing consultant in Rancho Rios, asked me about my research design. I told him I was interviewing both high- and low-wage workers to understand the role digital technologies played in inequalities in precarious work, and he looked disappointed. "Oh! I guess my interview was probably not that useful. I mean . . . I'm not rich or anything, but I'm not exactly struggling to pay my phone bill anymore, you know? You should've talked to me straight out of college [laughs]! I hope I didn't waste your time!" In his mind, Leonard's relative success and the ease with which he navigated his digital hustle meant that his experiences were less relevant to my understanding of digital inequalities and precarious work.

Leonard isn't alone in this logic. In studying digital inequalities, per-haps predictably, we haven't interrogated success and ease as much as we've investigated constraint and marginalization. As researchers, advocates, and activists concerned with the intersection of digital technologies and social inequalities, we have largely focused on the groups who are excluded from online spaces, who have to fight to be included, or for whom these tech-nologies have negative consequences to privacy, identity, and livelihood. However, we have mostly neglected to interrogate the ways that advantages accrue to those groups who benefit from these same systems.[16] Studies of marginalized tech users have pushed against universalizing accounts of technological progress and the democratizing potential of the Internet to give us more nuanced pictures of sociotechnical systems as they operate for those who were never imaged as users. As Tressie McMillan Cottom writes, "Nobody understands the motion of the ocean like the fish that must fight the current to swim upstream."[17] However, inequality is about relationships of power, and it's difficult to understand a relationship from only one side. Research that compares groups that occupy different positions within the same sociotechnical systems is essential if we are to fully investigate digital inequalities because understanding the construction of privilege helps us understand the social dynamics of marginalization.[18]

Comparative research in digital inequalities can surface similarities across disparate contexts. In Chapter 2, I observed that high- and low-wage gig workers shared not only the practices of the digital hustle but also a sense of pride and accomplishment in their ability to use their digital technologies to pull off the juggling acts required by independent and contingent work. While it's widely acknowledged that high-wage freelancers often choose the insecurity of independent work, in part, because of the autonomy and cre-ativity it offers, we don't often extend this privilege to low-wage workers, who are more often understood as being forced into less secure working arrangements. A comparative approach to digital inequalities can encourage us to examine the agency and choice vulnerable workers have in situations that we nevertheless recognize as constrained and exploitative. Instead of looking for different explanations for what may look like successful and constrained digital technology use, it's important to be able to see the possi-bility for shared motivations and meanings across disparate social contexts.

In the process of conducting this research, I've come to believe that a part of what reproduces and reinforces inequalities within the gig economy is that the digital hustles of workers at either end of the pay ladder are mostly invisible to one another. High-wage freelancers may feel they have little in common with workers struggling to piece together part-time work or find work through an app. The relationship between these two sets of workers is characterized by what Tara McPherson has called a "lenticular logic," a way of thinking that allows us to observe two things as happening at the same time but occludes or forbids us from understanding the relationship between them.[19] Public and policy discourse about the future of work creates a lenticular way of thinking about the fates of high- and low-wage workers; we recognize that highly educated, highly paid freelancers and contractors are a part of the same story as Instacart shoppers, but we can't see the conditions that bind them together. Accounts of inequality that place groups on either side of a divide may obscure similarities that could otherwise be the basis for future solidarities among independent and contingent workers.

Where Do We Go from Here?

Digital technologies will not solve poverty. Better digital skills and Wi-Fi hotspots will not end centuries of public choices that have undermined security and dignity for the many in favor of wealth for the few. Workers in this book are using digital technologies to fix problems that have been centuries in the making. More digital technologies will not fix these problems. These problems require structural solutions, including regulation that would transfer some of the risks of employment back to employers, and patch the gaping holes in our social safety net to ensure precarious work doesn't leave people to their own devices in the first place. However, there are also more immediate repairs that could be done to improve the lives of precarious workers.

The first is to consider more carefully how we achieve inclusion—for instance, by ending exploitative pricing and throttling of download and upload speeds based on how people are able to pay for their connectivity. Connectivity is essential to everything from finding a job to finding healthcare, and precarious workers are spending large portions of their fluctuating incomes on services to keep them connected. It's not fair that, once connected, they face poorer quality services. We can support organizations who're already fighting to end these practices, like the Movement Alliance Project (formerly the Media Mobilizing Project), and Free Press, among others.

Next, we can acknowledge the role of digital privilege by reframing digital literacy programs. Although high- and low-wage workers are using many of the same digital skills to find work, they're valued and supported in starkly unequal ways. Community organizations can reconsider the ways they understand constituents as already skilled and take a more collaborative approach to supporting their hustles through online labor markets. By offering help with small business services that high-wage workers are able to pay for, like branding, Web design, online retailing, identity protection, and digital security services, and also by convening community collectives to support resource-sharing among workers, community organizations can help redistribute digital privilege.[20]

Last, connectivity is a nonstop balancing act for many precarious workers. Employers should stop assuming unproblematic universal access is the norm. From large employers down to household employers, if you expect and enjoy the flexibility that comes from on-the-go connectivity from the people who work for you, take steps to subsidize it. Of course, these costs shouldn't only be borne by employers, and because many of these workers don't have a single employer—or any formal employers—we also need to think about ways of sharing or distributing these subsidies among those who require workers to be connected.

I think these steps could help support workers who shoulder the burdens of a polarized and increasingly digitized market for precarious labor. Of course, they're limited in their ability to address the structural causes, and not only the symptoms, of these ways of working. It's important to remember that the digital hustle is both a solution to these deeper problems and a symptom. The precarious workers in this book use their technologies to create livelihoods in tough circumstances, but we also need to recognize how these very tools are at the center of an ideology of personal responsibility for economic mobility and opportunity that has undercut possibilities for higher wages, more stable benefits, and a stronger and broader social safety net. In other words, it's important to recognize that these workers are indeed strong and resilient and deserve any and all immediate steps we can take to bolster these admirable qualities. However, it's also important to recognize that they cultivate these qualities within a system that makes personal resilience the only path to survival. These are the roots of the system that need to be addressed, the one that leaves workers to their own devices in the first place.

As we stare down the uncertain future that will emerge in the wake of the COVID-19 crisis and its consequences for workers at either end of the income ladder, it's more important than ever to understand digital inequalities in ways that allow us to see the connections between very different experiences of our increasingly intimate relationships with digital technologies. This pandemic has created profoundly isolating and fragmented daily experiences of work and use of digital technologies. High- and low-wage workers face what feels like entirely different crises, with workers on one side buying better home office equipment and workers on the other side making tradeoffs between income and safety for themselves and their families by logging into an app to find work or navigating glitchy and overloaded online systems to secure unemployment benefits or other kinds of financial relief. We need to reckon with the hidden similarities and uncomfortable differences in our experiences with sociotechnical systems to understand the current role of digital technologies in our working lives and to build a more equitable economy for the future.

Methodological Appendix

This appendix is a reflection on the process of conducting ethnographic interviews with workers across the United States about their work and their digital technologies. In the tradition of ethnographers across the social sciences, I feel it's important to explain both the methodological decisions and definitions used in this research as well as reflect on the ways my presence and perspective shaped the process of research. The first section reflects on my social position, both inside and outside the frame of the research process, and the rest explains my research design, sampling, and operationalization of key variables. I also discuss the practices of ethnographic interviewing, my process for analyzing them, and the limitations of the research.

Sitting Across the Table: Social Position and the Researcher's Role in Interpretation

We're all lay ethnographers and interviewers, making observations and hypotheses about the people around us. As a qualitative researcher, I do that with the people who participate in my research, but I am also a participant in the sensemaking of others. Donna Haraway warns that "claiming the power to see and not be seen, to represent while escaping representation" is to deny that all knowledge comes from somewhere.

As a scholar, I regard myself as a qualitative social scientist interested in issues of identity, dignity, and digital inequalities, and I carried out this fieldwork as a PhD student in sociology from 2013 to 2016. I follow in the footsteps of qualitative scholars like Barbara Ehrenreich, Annette Lareau, Arlie Hochschild, and Kathryn Edin, who also seek to understand the ways that disadvantages are perpetuated, felt, and understood by taking the words and experiences of economically marginalized people seriously. I also follow scholars like danah boyd, Sherry Turkle, and Judy Wajcman in their investigations of the social and cultural lives of digital technologies. And, as a White woman, I also recognize that I share more with these women than simply my social scientific training; a position of privilege is afforded to me as a White woman studying people who have been systematically and explicitly excluded from the same institutions that trained me and now support my research.

When I started this research project, I didn't fully understand the ways that racism shaped the politics swirling around technology and work. In 2017, as I watched the incomprehensible violence and hatred of the Unite the Right Rally ravage the city and residents of Charlottesville, Virginia, and afterward as I learned about the role of social media and other digital information and communication technologies in that event and the many that followed, I realized that these were not issues outside of the questions that kept me up at night and certainly not things I could safely ignore because I didn't "study race." I have and continue to struggle with the ways my social position shapes not only my observations of the social world and the things participants choose to share with me in interviews but also the very questions I ask in the first place.

My own social position shaped what my interviewees chose to tell me and, inevitably, how I heard and made sense of their stories. Unlike traditional ethnography, where researchers gain entry and, hopefully, trust within a community and observe everyday interactions over longer periods of time, ethnographic interviewing entails more fleeting interactions with participants. This requires that I establish rapport with interviewees relatively quickly, leading me to pay attention to signals from my participants about the way they see me and include these moments in my fieldnotes.

During interviews, I found the people I spoke with generally forthcoming and open about the work they did and how they used their digital technologies, but there were times when my position as a relatively young and presumably upwardly mobile student may have led my interviewees to feel uncomfortable sharing their stories of downward mobility and struggles to find work. Upon asking Charlie, a self-employed home contractor in Washington, DC, what I thought was a straightforward question about whether he worked full-time or part-time, he answered, "Because of the economy falling down the tubes, it's become between full-time and part-time, it's like middle-time. I'm still my own boss and everything, and I'm still making it work, but, you know, it's [pause] you know I've been trying to branch out and do other things, too." Charlie buttressed his acknowledgment of his lower income with a reminder about his autonomy and independence as his own boss. These experiences and others sensitized me to the ways that my interviewees may have viewed me and wanted me to view them.

A few weeks later, as a low-wage participant in Washington, DC, was putting his phone away and gathering his things after a long and deeply personal interview, he asked if I was a social worker. I repeated my explanation about being a graduate student and asked what made him think that. He paused and looked a little uncomfortable before saying, "Well . . . and I don't mean any offense by this, but you seem like a nice White lady who's worried about other people's problems."

Catching glimpses of myself through my participants' eyes didn't just happen with the low-wage workers in this study. At the beginning and end of every interview, I asked participants if they had any questions for me, and before my interview with Grace, a Washington, DC–based consultant in her early forties, could even begin, she peppered me with questions about my career ambitions, asking me what I wanted to be "when I grew up" with a half-joking smile. Over the din of the happy hour crowd around us, I shared my ambitions to be a professor, and she replied, "Awww! You're so cute!" I present these moments not because they're about my interviewees but because they're about me. As a qualitative social scientist, I set out to try to understand the world through the eyes of my participants and can only do so when I recognize my own role as a participant in that very process.

I'm no stranger to low-wage work, having slung Subway sandwiches and worked at food courts throughout high school. In college, I received federal work-study assistance for greeting visitors at my college's information booth and supplemented those wages by working as a "brand representative" for different beer and liquor companies, pushing samples and T-shirts in bars and liquor stores. Many of my friends and family still work at jobs like these. However, having had the privilege to finish college and graduate school with people on the path to high-status professions and lucrative careers, I'm particularly attuned to the stark differences in rhythm and narratives of work between these two worlds. While this history was, I believe, invisible to my interviewees, it nonetheless shaped the themes and feelings I encountered in analyzing their words and writing these chapters.

Research Design

For this study, I wanted to take a perspective that went beyond on-demand work done through online platforms. With the rise of Uber, Lyft, TaskRabbit, and other companies, I was intrigued by the way the world's attention started to turn to the importance of digital technologies to forms of low-wage work that hadn't *appeared* to be transformed by the technologies that were now essential to white-collar work. As I show in this book, digital technologies were already becoming essential to these kinds of low-wage jobs, and I was frustrated that this only seemed to further marginalize these workers and their issues from public conversations about the future of work. Instead of sampling on the dependent variable and studying workers using a particular online platform or a workplace that was undergoing technological change, I decided to take a cross-sectional and comparative approach to independent and contingent work. By not limiting the types of occupations and jobs that participants had (e.g., retail sales, manual labor), I was able to gain a cross-sectional view of the wide variety of practices operating across many different types of contingent labor markets. I opted instead to constrain the sample by work arrangement rather than type of work (see later discussion on sampling).

Additionally, a comparative design allowed me to contextualize workers' experiences across the class divide. In research examining "divides" in technology use, scholars like Susan Herring and danah boyd have pointed out that researchers risk exoticizing the lived experiences of those with very different relationships to technology and media, describing them as alien from their own concerns and experiences, when really, their motivations—to build friendships, construct identity, or even just to stave off boredom—may actually be quite similar to their own. A comparative approach allowed me to situate high- and low-wage workers' practices in their own unique contexts but also avoid attributing patterns within each group to class or occupational status, when, as the book illustrates, the conditions of contingent work itself pattern digital technology practices across classes. I address the necessity of comparative research in understanding digital inequalities more extensively in the concluding chapter of this book.

Sample

We tend to talk about the gig economy as if it's a separate and distinct part of the labor market, but there are substantial debates about how to think about the types of jobs and tasks that have become associated with this kind of work. In 2017, the Bureau of Labor Statistics (BLS), the part of the US federal government tasked with taking the temperature of the US workforce, wanted to count the number of US workers who were providing in-person services through platforms like Uber and doing "micro-work" tasks, oftentimes at home, through platforms like Amazon Mechanical Turk. They asked thousands of people a series of questions about whether or not they did "electronically mediated work." After the surveys came back, the BLS was surprised when they found that doctors, police officers, and other professionals had indicated their work was electronically mediated. In a press release, the BLS said that they had to "recode erroneous answers," and that it was "difficult" to isolate the workers they were interested in because many people interpreted the question as asking whether they used a computer or mobile app to do their jobs.[1] Although they saw it as an error, the BLS had stumbled upon the growing reality that very few of our jobs nowadays could not be described, at least in part, as "electronically mediated." Digital technologies have become a defining piece of many different kinds of work, from low- to high-wage, manual to professional, precarious to secure.

Among labor scholars, distinguishing gig work from traditional work is usually not a question of technology but instead turns on differences in the relationship between workers and employers. However, some scholars have gone further to define broader transitions in job quality that are linked to these changes to the employment relationship. Arne Kalleberg defines precarious work as "uncertain, unpredictable, and risky from the point of view of the worker."[2] However, the BLS takes a more neutral and narrow approach by defining both "contingent" and "alternative" work. For the BLS, contingent workers are those who don't expect their jobs to last or those who don't have "implicit or explicit contracts for ongoing employment," for example, seasonal sales associates, while those with "alternative employment arrangements" are comprised of "independent contractors, on-call workers, temporary help agency workers, and workers provided by contract firms."[3] I used a combination of these two definitions to formulate a series of screening questions, including asking if the person worked for pay at least part-time and the number and type of different jobs that generated income for them, including a sense of what they considered their "main" type of work and/or source of income.[4] When I met workers, in-person or online (see more on locations of recruiting later), I would ask them the screening questions and obtain some basic contact information to follow up later. To qualify for the study, participants had to be working at least part-time, defined as twenty hours per week, in some kind of independent gig work.

Despite more widespread practices of remote work and the association between digital technologies and the ability to work from anywhere, the kinds of independent and entrepreneurial work and Internet connectivity available to workers in the United States are still influenced by their location. Only about 5 percent of Americans report that their jobs are based entirely at home, and these practices are concentrated among the most educated workers.[5] Geographic location still influences the kinds of work available and its organization, with particularly strong differences between urban and rural areas. Differences in Internet access between rural and urban areas are also pronounced. For example, 68.6 percent of rural Americans have access to both high-speed home broadband and mobile Internet, as opposed to 97.9 percent of Americans in urban areas.[6]

To capture some amount of this variation, I conducted interviews in four cities. Washington, DC, was my primary urban fieldsite, supplemented by a few interviews done in New York City as the city has become a central hub of the on-demand economy.[7] I also interviewed in a mid-sized, affluent suburb of San Francisco that I call Rancho Rio and a small, rural town outside of Buffalo, New York, I call Mainville. I decided to give the suburban and rural fieldsites pseudonyms to protect respondents' anonymity but felt this was unnecessary in the larger urban sites. See Table A.1 for more descriptive information about each of these sites.

In these cities, I used publicly available demographic information and relied on local guides to choose neighborhoods where high- and low-wage workers might live and scouted those neighborhoods for the retail stores that sold digital information and communication technologies. Traditionally, qualitative studies of work have focused on workplaces or other physical locations where the workers they were interested in gathered. However, the workers I was interested in don't have single workplaces but are instead constantly moving between clients, employers, and even types of gigs, sometimes all in the same day. For these reasons, I chose to focus on recruiting in retail locations where these workers purchase and service their digital technologies.

Table A.1 Descriptive Statistics for Fieldsites

	Washington, DC	Rancho Rio, CA	Mainville, NY	New York, NY	All United States
Population	647,484	72,203	11,761	8,426,743	
Economic Indicators (%)					
Unemployment	6.4	5	8	5	4.2
Poverty	18	4.7	31.1	20.6	12.7
College degree +	54.6	47.8	14.6	35.7	33
Median HH income	$70,848	100,163	$34,856	$53,373	$59,039

Source: American Community Survey, 2015, Bureau of Labor Statistics, 2017.

Consumer electronics retail is highly segregated by class. Major mobile phone carriers often segment their brands that cater to low-wage consumers to avoid association with their main brand identity. In Mainville, New York, there were only three stores that sold cell phones in town (one of which was a convenience store) and only one (Walmart) that also sold computers; all catered to the town's mostly low-income residents (see Chapter 3 for more on these locations). I learned from hanging out at these stores that the town's wealthier residents drove between twenty and forty-five minutes to a Verizon or Apple store in slightly larger cities, so I also recruited at a widely attended holiday parade in town.[8] To ensure I was reaching people with different shift schedules, I would vary the hours I spent at retail stores, including weekends. In the suburban fieldsite, I recruited at a Best Buy and a Verizon store over the course of ten days.

In my urban fieldsite, I recruited at Apple, MetroPCS, Cricket Wireless, and several "side stores" that sold phones alongside snacks, videogames, jewelry, and household goods.[9] It proved difficult to recruit high-wage independent workers at retail stores, and early interviews suggested that they tended to avoid spending time traveling to or waiting in retail stores by contracting out their IT work or ordering online, leading me to also recruit online. I posted on Facebook groups associated with my own and others' colleges and universities. To ensure I hadn't introduced a systematic bias by using this Internet-based recruitment technique to find high-wage workers in Washington, DC, I used online postings to recruit low-wage workers as well. After learning from early interviewees about the importance of Craigslist to find gigs, I used online postings in the gigs section of this site in Washington, DC. I did not find that interviewees recruited online varied in any substantive ways from those recruited in retail stores.

Operationalization of Wage and Technology

Much like the definition of contingent work itself, there are a number of ways that I could have divided workers into descriptive categories that describe their type of work. Many advocate for differentiating along the lines of autonomy and control in combination with wages.[10] The issue of control is complex in the case of contingent work and can shift from gig to gig. Also, the use of digital technologies is wrapped up in workers' sense of control over their work.[11] After initial interviews, it appeared that workers' self-reported wages were a parsimonious way to sort workers into categories of high- and low-wage groups. Throughout the chapters, I use wage instead of class or "income" because I felt that this was closer to the way workers themselves made sense of their earning power. For contingent workers at either end of the class divide, income instability is a constant concern. Respondents with their own business licenses and contract IT help would talk about constructing their annual budgets around slow and busy months or seasons in ways that were not that far removed from landscapers who were moonlighting as seasonal sales associates in the winter months. Income varied substantially from month to month for these workers, but their hourly rates, either the rates they charged clients or the wages they looked to get from employers were a more stable and meaningful measure of their work.

To formulate categories of high- and low-wage workers, I used census data about median hourly wage in the United States (just over $17). Mostly, participants sorted into well above and well below this number, but for those who fell closer to the median, I used a combination of education and occupational status to decide which category to put them in. Even within such a straightforward measure, there were complexities and contradictions that were simplified into these dichotomous categories. See Table A.2 for descriptive statistics about the sample.

This study did not limit the kinds of digital technologies included in interviews and instead used an inductive definition based on participants' own understandings. The use of digital technologies was not a part of my screening questions to qualify participants for the study. Oftentimes, I had to actively work against the tendency of people to discount their own technical practices and skills before I could even ask the screening questions. Before I asked the questions, as I was briefly describing who I was and what I was doing, I would explain I was interested in how people used digital technologies for their jobs. This description of the study would sometimes result in potential interviewees insisting that they "didn't know how to use all that stuff," or that they could "barely use"

Table A.2. Interviewee Job Characteristics

	Washington, DC	Rancho Rio, CA	Mainville, NY	New York, NY	Total
Wage Category					
Low	20	5	19	10	54
High	20	10	14	2	46
Characteristics					
Full-time	31	12	22	3	68
Multiple jobs	6	2	4	3	
Part-time	9	3	11	9	32
Multiple jobs	2	2	8	9	
Self-employed	15	10	13	5	43

Source: Self-reported.

their phones.[12] When this happened, I'd generally asked them if they had a cell phone or a Facebook account and if they ever used it in relation to their work; when they'd usually nod, I'd respond that I'd love to hear about it and press on with my screening questions. At the start of the interview, after asking participants to detail all the work they did for pay, I'd ask them to list the digital technologies they used in connection with those jobs. Despite the broad latitude this approach gave interviewees, there was considerable overlap in the technologies they listed and discussed in their interviews (see Fig. A.1). These included pieces of hardware or devices, features of devices (e.g., calling, texting), and applications or websites (e.g., online social networks, scheduling applications).

Ethnographic Interviewing

I conducted interviews with one hundred high- and low-wage independent and contingent workers across many different industries. I used ethnographic interviewing, which refers both to the techniques the interviewer uses during the interview itself and the way the interview is understood as a unique kind of social interaction. Drawing on the traditions of ethnographic research from anthropology and sociology, ethnographic interviewing treats interviewees as "informants," understanding them as experts in their own experiences and inviting them to teach the interviewer about the world as they know it.[13] This method also understands interviews as a kind of account-giving, meaning that the interviewer isn't only interested in the content of the answer but also the way answers are given. Ethnographic interviewers assume that answers to their questions aren't a straightforward reporting of facts as they happened to the participant but are meaningful accounts that emerge from the standpoint of the participant in the social context of talking to a researcher.

I also kept detailed ethnographic fieldnotes about the homes, coffee shops, fast-food joints, and libraries where I interviewed participants and about the interactions I observed and participated in during the process of recruitment and interviewing. These observations are used throughout this book to contextualize my interpretation of the interviews and participants' stories. Interviews were conducted mostly one on one and face to face and were audio-recorded and transcribed. Interviews were semistructured and ranged in length from forty-five minutes to an hour and thirty-seven minutes. I asked interviewees a series of questions about their digital technologies and gigs, which ones they used regularly in the course of their work, and how they used them. I asked them to tell me stories about times when coworkers, bosses, or clients used technologies in ways that annoyed them, stories about when friends and family members had tried to contact them while working, and times when they felt they had annoyed others with their technology practices. I also asked several broader opinion questions on the general relationship between technology, work, and society today. These questions were designed to elicit their everyday practices with their digital technologies without asking about specific practices (e.g., scheduling, client communication, etc.), as well as inductively derive their normative frameworks for evaluating proper and improper or professional and unprofessional technology practices. I used a standard set of interview questions for all participants and asked follow-up or probing questions depending on the stories they shared. I asked participants to choose the location of our interview, and I would show up early to observe the location and choose a seat and stay after the interview was over to record fieldnotes about the interview and location.[14] After listening to a description of the study and obtaining their initial written consent, participants received $20 cash before we began the interview and were told they could

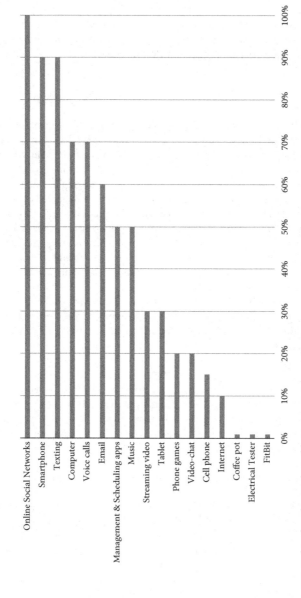

Fig. A.1 Digital technologies listed by interviewees, all sites.

stop at any time and leave with the $20, which nobody did. After the interview, I gave each participant a survey to obtain basic demographic and wage information.

Data Analysis

Once all interviews were transcribed, my analysis of these texts followed several different stages of coding for clusters and categories of practices and the feelings that accompanied them as well as the judgments interviewees leveled at others in their social worlds. I followed a grounded theory approach, but as coding progressed, I created a codebook that I used to code subsequent interviews and recode earlier interviews to ensure consistency across the sample.[15] As I prepared an analysis of a smaller subset of these interviewees for a publication, I gave a set of codes to a group of fellow sociologists to check against my own interpretations and provide feedback. I also checked my interpretations and solicited feedback on the terms I was developing with several participants in both high- and low-wage groups throughout the process of analysis.

There were often multiple intersecting axes of power and privilege that shaped interviewees' experiences, and I returned to the transcripts many times in an attempt to narrow an analytic spotlight onto the axes that were "more present, more apt, or more explanatory than others."[16] Occasionally, in this book, I combined descriptions of multiple individual participants into a single person to prevent reidentification of individual participants. Inevitably, factors that may have explained some variation in people's practices and feelings are left out of this book, and the narrative I tell about their experiences is incomplete. However, I have striven to honor the time and stories entrusted to me by the people who were generous enough with their time and giving enough in their spirit to chat with me about their lives for a while.

Limitations

As with all research, the findings presented in this book are limited in some important ways. This study relied on participants' reports of their own practices and didn't employ ethnographic methods that may have allowed me to check their reports against their behaviors or corroborate their accounts with those of their managers, employers, and clients. Interviews are limited in the kind of information they can solicit from interviewees, as people's accounts don't always line up with what they actually do.[17] It's possible that participants gave accounts of their digital hustling that were embellished or otherwise inaccurate. However, the aims of this study were not only to document what people do with their digital technologies but also what those practices meant to them. As Allison Pugh points out, interviews are uniquely situated to "access different levels of information about people's motivation, beliefs, meanings, feelings and practices—in other words, the culture they use" to account for their daily lives.[18]

In using a cross-sectional design rather than one more focused on a particular kind of occupation, it's possible that I missed types of digital technologies and practices that are important to some types of work but weren't mentioned by interviewees. This decision also meant that I only caught glimpses of technological systems that were deeply important—but only for a few workers (e.g., enterprise scheduling software, online labor platforms). There were also forms of demographic diversity that may have influenced both work and technology practices that were underrepresented in my sample. I didn't sample in ways that would allow me to see variations patterned by sexuality, disability, gender identity, or immigration status.

Finally, I began interviews for this project in 2013, before Cambridge Analytica, the Unite the Right rally, and a broader public reckoning with the role of digital technologies in surveillance, algorithmic discrimination, and networked hatred. Some important things have changed in the universe of digital technology and media that surrounds us, but many things haven't. It's likely that my interviewees may be using different technologies and would give different answers today than they did several years ago, but in my ongoing conversations with them, the practices outlined here have only deepened in intensity and widened in range. The issue of the age of this data is a nontrivial one but one that I believe only raises the stakes of the findings presented here and makes it all the more urgent to understand the ways social inequalities shape our approaches to our work and the technology we increasingly depend on to do it.

Notes

Introduction

1. In their accounts of post-Fordism, postindustrialism, and the rise of the "network society," social theorists contemplate the fate of work under new technological conditions. See Drucker, *Landmarks of Tomorrow*; Bell, *The Coming of Post-Industrial Society*; Castells, *The Rise of Network Society*; Kumar, *From Post-Industrial to Post-Modern Society*; and Zuboff, *In the Age of the Smart Machine*.
2. For research that investigates how white-collar workers are affected by these technological changes, see Chesley, *Blurring Boundaries?*; Chesley, *Families in a High-Tech Age*; Duxbury et al., *From 9 to 5 to 24/7*; Moen et al., "Overworked Professionals"; Perlow, *Boundary Control*; Perlow, *Sleeping with Your Smartphone*; Schieman, Milkie, and Glavin, *When Work Interferes with Life*; Turkle, *Alone Together*.
3. Smith, *Gig Work*.
4. For more on this definition, see the methodological appendix.
5. Drawing on the work of Marcel Mauss, Ashley Mears points out that when work escapes the traditional workplace, a relational, rather than place-based, framework for understanding the production of labor's value is necessary: "when conceptualized in relational rather than physical space, the value of labor emerges through personal ties and webs of reciprocity—the very heart of all economic exchange" (Mears, "Working for Free," 1100). This intervention also suggests that the technologies that facilitate these relational ties and webs are all the more important to understand now that work is changing.
6. On morality tales of consumption for the poor, see Pugh, *Longing and Belonging*. On Chaffetz, see Scott, "Chaffetz Walks Back." On the important role of mobile phones in coordinating health care, see Gonzales, Ems, and Suri, "Cellphone Disconnection Disrupts."
7. As Susan Leigh Star explains, infrastructure is "the forgotten, the background, the frozen-in-place." Star, "Ethnography of Infrastructure," 379.
8. Parks, "Antenna Tree."
9. Ravenelle, *Hustle and Gig*; Schor, *After the Gig*.
10. Weil, *Fissured Workplace*; Hyman, *Temp*.
11. On flexibility and casualization, see Hollister, *Employment Stability*; Kalleberg, *Good Jobs*; Smith, *Crossing the Great Divide*.
12. Hacker, *Great Risk Shift*.
13. Farrell et al., "Online Platform Economy." In 2016, Pew put their estimate of Americans who had earned money doing tasks through platforms at closer to 5 percent. Smith, *Gig Work*. The Bureau of Labor Statistics (BLS) defines contingent workers as those

"who do not expect their jobs to last or who report that their jobs are temporary. They do not have an implicit or explicit contract for continuing employment" (Kosanovich, *A Look at Contingent Workers*). Other scholars use a wider-reaching definition that includes workers in many types of "nonstandard" or "alternative" employment arrangements, which includes independent contractors, part-time workers, and contract company employment. The BLS estimate includes workers in both of these categories. This book uses the term in the second sense. The Government Accountability Office estimates of this population are based on data from the General Social Survey, using a broader definition of alternative employment that included standard part-time workers and day laborers. Kalleberg, "Nonstandard Employment Relations"; Government Accountability Office, *Contingent Workforce*.

14. Between 2013 and 2019, the percentage of Americans subscribed to home broadband only increased by 3 percent. Americans' rates of adoption of laptop computers declined between 2010 and 2015. More Americans report going online using their smartphones than their laptops or desktop. Pew Research Center, "Demographics"; Pew Research Center, *Mobile Technology*; Anderson, *Technology Device Ownership*.

15. Seventeen percent of Americans own a smartphone to go online and don't subscribe to home broadband. Pew Research Center, *Mobile Technology*.

16. Blumberg and Luke, "Wireless Substitution."

17. Matsakis, "Carpenter v. United States"; Ticona and Selbst, "In Carpenter Case."

18. Though 17 percent of Americans aged 18–29 are smartphone dependent, this general statistic obscures dramatic differences between different racial, ethnic, and socioeconomic groups: 23 percent of Hispanic people and 15 percent of Black people (compared to 9 percent of White people), 21 percent of people making less than $30,000 (compared to 5 percent of people making over $75,000). Madden et al. found even starker divides, "63 percent of smartphone Internet users who live in households earning less than $20,000 per year say they mostly go online using their cell phone, compared with just 21 percent of those in households earning $100,000 or more per year" (70). In addition, "Adults living in poverty (66.3 percent) and near poverty (59.0 percent) were more likely than higher income adults (48.5 percent) to be living in households with only wireless telephones." See Blumberg and Luke, "Wireless Substitution," 3. See also Mossberger, Tolbert, and Anderson, "The Mobile Internet." Pew Research Center, "Mobile Device Ownership"; Madden et al., "Privacy, Poverty."

19. While more than half (55 percent) of Black and Hispanic smartphone owners use their phone for job-related activities, only about a third (37 percent) of White people do. Black and Hispanic people are also twice as likely to use their phone to apply for a job as White people. While these racial and ethnic differences cut across class, minorities from low-income households are the most likely to be heavily dependent on their smartphones for access to the Internet. Nationwide, 13 percent of low-income Americans, 12 percent of African Americans, and 13 percent of Latinos depend on their mobile devices for access. Smith, "Searching for Work." While poor households are less likely to have smartphones overall (60 percent of households below poverty level have a smartphone versus 77 percent of those above poverty line), when they do,

they're more likely to rely on that phone for Internet access (15 percent of households below the poverty line rely on them versus 6 percent of those above poverty line). The dynamics are the same when looking at employed and unemployed people—while 76 percent of unemployed people have smartphones (versus 86 percent of employed people), 14.7 percent rely on them for Internet, compared to 7.9 percent of employed people. Madden found even starker patterns, with 63 percent of smartphone users living in households earning less than $20,000 a year reporting they mostly go online using their cell phone compared to just 21 percent of users in householders making $100,000 or over. People from lower income households are especially likely to use their phones in their job searches; while 43 percent of smartphone users report using them to look up information about jobs, users from households that make less than $30,000 a year are twice as likely as those from households that make over $75,000 a year to do so. Lewis, *Handheld Device Ownership*; Madden, *Privacy, Security*; Smith, *Searching for Work*.

20. As Wendy Chun explains, "media matter the most when they seem not to matter at all." Chun, *Updating to Remain*, 1. See also Ling, *Taken for Grantedness*.

21. More recently, scholars have turned away from the idea of the digital divide and toward the term "digital inequalities" to examine how social marginalization shapes digital technology use. However, the exclusion paradigm is still predominant. See also Helsper, "The Social Relativity"; Jenkins, "Cyberspace and Race"; Rodino-Colocino, "Laboring Under"; and Selwyn, "Reconsidering Political" for critiques of the "digital divide" as an analytical framework.

22. Stuart (*Ballad of the Bullet*, 9) points out that "long-standing inequalities shape the ways outside parties read and react to different users of technology." These reactions are an essential piece of studying what he calls "digital disadvantage," or the ways "different people, with contrasting levels of privilege, fatefully engage with the same technologies in their daily lives" (8).

23. Wilson, *When Work Disappears*.

24. Greene, *Promise of Access*.

25. The use of the digital divide frame has fallen out of favor with researchers. For more comprehensive critiques and reconceptualizations of the idea of the digital divide, see Dimaggio et al., "From Unequal Access"; Gonzales, "The Contemporary US Digital Divide"; Hargittai, "Second Level Digital Divides"; Hargittai, "Digital Na(t)ives?"; Greene, "Discovering the Divide"; Selwyn, "Reconsidering Political"; and Warschauer, *Technology and Social Inclusion*.

26. On divides in access, see Waschauer, *Technology and Social Inclusion*; and van Dijk, *Deepening Divide*. On divides in skills, see van Deursen and van Dijk, "Digital Divide Shifts"; and Scheerder et al., "Determinants of Internet Skills."

27. As this area of study has expanded, researchers first refined several different types of divides and then undertook a more thoroughgoing redefinition of this field as the study of digital inequalities rather than divides. Robinson et al., "Digital Inequalities"; Dimaggio et al., "Digital Inequality"; Helsper, "Social Relativity."

28. Castells, *Internet Galaxy*, 3.

29. Labaton, "New F.C.C. Chief."

30. Lifeline expanded to include mobile phone subsidies and was codified under President Bill Clinton in 1996, but, recognizing the importance of mobile Internet access, the first-term Obama administration expanded the program to allow recipients to choose a wired phone, mobile phone, or broadband subsidy of $9.25. In the aftermath of Hurricane Katrina, President Bush used the Universal Service Fund (a fund administered through mobile providers and passed onto customers, the same fund used to fund the Lifeline program) to provide free prepaid wireless airtime so that those displaced by the storm could stay in touch. He later expanded the program beyond those affected by Katrina. Ninety-seven percent of Lifeline participants qualified because they were already participating in another type of public benefits, with Supplemental Assistance Nutrition Program (SNAP) or Medicaid being the most common (making up 68 percent of participants). As of this writing, a Lifeline applicant can qualify by having an income at or below federal poverty guidelines or by participation in another federal assistance program (such as SNAP, SSI, Medicaid). Only one subsidy (toward one phone or toward broadband subscription) per household is allowed. Anon, "Lifeline Program Statistics." In 2017, over 90 percent of people used the $9.25 per month subsidy toward mobile phone service. Eligible households get one subsidy, paid directly to the phone company or Internet carrier, meaning that households need to choose between phone and broadband service and, depending on the carrier, usually receive one phone. Through Assurance Wireless, a Lifeline-only sub-brand of T-Mobile, a Lifeline-eligible household in Pennsylvania could receive one free low-end Android smartphone, 2 GB of free data per month, free unlimited texting, and 350 free minutes of voice calls.

31. Anon, "What We're Up Against"; Anon, "Obama Donors." One thread on Stormfront discussing the video has more than 4,000 posts.

32. The welfare queen trope emerged during 1970s and was used by President Reagan's 1976 campaign to galvanize support for proposed cuts to welfare programs. The specter of welfare fraud and the demonization of Black women as its supposed perpetrators has reemerged many times in political projects aimed at cutting public entitlement programs. See Gilliom, *Overseers of the Poor*.

33. The meaning of digital inclusion is not outside the political economy of technology but central to our understandings of it. As Eva Illouz points out, "meanings serve as weapons in the struggles of social groups to secure and further their interests. . . . Culture thus is a matter of shared meanings, but it is not only that: it is also one of the ways in which exclusion, inequality, and power structures are maintained and reproduced." In these ways, struggles over the meaning of digital technologies as luxuries or as necessities are "proxy wars" in larger struggles over the best way to ensure economic equity. Illouz, *Consuming the Romantic Utopia*, 6.

34. Cooper, *Cut Adrift*; Hochschild, *Their Own Land*; Pugh, *Tumbleweed Society*; Silva, *Coming Up Short*.

35. Gregg, *Work's Intimacy*; Neff, *Venture Labor*; Sennett, *Corrosion of Character*.

36. Petriglieri et al., "Agony and Ecstasy."

37. Lane, *A Company of One*; Liu, *Laws of Cool*.

38. Turner, *From Counterculture to Cyberculture*.

39. Negroponte, *Being Digital*.
40. Gregg, *Work's Intimacy*.
41. Anon, "Managing the Future."
42. Becker, "Arts and Crafts."
43. Hodson, *Dignity at Work*.
44. Sennett, *The Craftsman*, 25.
45. Braverman, *Labor and Monopoly Capital*; Ritzer, *McDonaldization of Society*.
46. Horowitz et al., *Trends in US Income*.
47. While much electronics retail has moved online, most smartphone sales (88 percent) still happen in physical retail stores. Smartphones also remain the most widely used form of digital information and communication technology in the United States, with 88 percent of Americans owning one, compared to 74 percent of people who own laptops/desktops, and smaller percentages owning tablets and e-readers. However, after finding from initial interviews with high-wage workers that much of their purchasing and shopping happened online, I started to recruit high-wage workers through social media. I replicated this technique among the low-wage workers using Craigslist and other online forums. See the methodological appendix for more detailed information on recruitment. Klaehne, "Amazon Leads"; Pew Research Center, *Mobile Fact Sheet 2019*.
48. T-Mobile, Sprint, and AT&T all offer prepaid phones and services under different brand names to consumers with low incomes and poor or little credit. For instance, Cricket Wireless, which offers mainly prepaid services, is owned by AT&T. Boost Mobile is owned by Sprint. In 2013, T-Mobile bought MetroPCS and rebranded it as Metro by T-Mobile and, at the time of this writing, is brokering a merger with Sprint. Verizon Wireless is the only national carrier that doesn't segment its low-cost wireless services under a different brand name. In addition, companies like Straight Talk, Tracfone, and Assurance Wireless, which offer exclusively prepaid services catering to low-income consumers, lease access from national carriers to serve their lower income consumer base.
49. My understanding of social classification and the reproduction of social inequalities draws on Collins's (1990) concept of the "matrix of domination." Other theorists of social class and its relationship to work have pointed to the enduring importance of race and gender in defining social class relationally or based on the amount of control workers can exercise at work. Collins, *Black Feminist Thought*; Wright et al., "The American Class Structure."
50. Bonilla-Silva, "Rethinking Racism"; Omi and Winant, *Racial Formation*.
51. Institutions, interactions, and audiences all shape how different social categories become salient for understanding people's use of digital technologies at different times. West and Fenstermaker, "Doing Difference"; Wimmer, "Making and Unmaking."
52. See Beck and Beck-Gernsheim, *Individualization*.
53. Swidler, "Culture in Action."
54. In this book, the actions that I've termed "strategies" would be closer to de Certeau's definition of "tactics," or the ways individuals improvise to navigate institutionalized rules in ways that are shaped but not determined by existing arrangements of power. De Certeau, *Practice of Everyday Life*.

Chapter 1

1. Orlikowski and Scott, "The Algorithm and the Crowd."
2. Foucault, *Discipline and Punish*; Christin, "Counting Clicks"; Rosenblat and Stark, "Algorithmic Labor and Information Asymmetries."
3. Espeland and Sauder, "Rankings and Reactivity."
4. Marwick, *Status Update*, 166.
5. Eubanks, *Digital Dead End*.
6. Neff, *Venture Labor*.
7. With contingent and flexible schedules and revolving doors of shortened or temporary contracts, workers spend a lot of time in a work frame of mind, whether it be looking for work or negotiating schedules, without being paid for it. Snyder calls this "work-for-labor" (*The Disrupted Workplace*, 212). On unpaid work done to prepare for paid labor, see Williams and Connell, "Looking Good and Sounding Right"; and Avery and Crain, *Branded*.
8. Arcy, "Emotion Work"; Duffy, *(Not) Getting Paid*; Jarrett, *Feminism, Labour, and Digital Media*; Terranova, "Free Labor."
9. Shade, "What the Hustle Looks Like."
10. Spence, *Knocking the Hustle*, 2.
11. My use of the term "hustle" differs from its use in other social science accounts. From studies of gangs (Short and Strodtbeck, "Group Processes") to drug dealers (Adler and Adler, "Shifts and Oscillations") to con artists (Grazian, *On the Make*), researchers have pointed out how disadvantaged people, "rather than challenge the system or push their way into the mainstream, organize to exploit it from the edge. Usually this means exploiting those who are more vulnerable" (Schwalbe et al., "Generic Processes," 429). While the digital hustle doesn't entail the exploitation of vulnerability or illegal activity, both high-and low-wage workers accept conventional goals but not the conventional means to achieve them (Merton, *Social Theory and Social Structure*; Cloward, "Illegitimate Means").
12. Wacquant, "Inside the Zone," 4.
13. Wacquant, "Inside the Zone," 4.
14. Temporary, under-the-table gigs are crucial in high-poverty neighborhoods and oftentimes the primary, if not only, way people engage in paid work. See Sugie, "Work as Foraging"; Edin, Lein, and Jenks, *Making Ends Meet*; Edin and Nelson, *Doing the Best I Can*; and Venkatesh, *Off the Books*; Edin and Shaefer, *$2.00 a Day*.
15. Katz and Krueger found that alternative work arrangements had risen by 50 percent and accounted for "94 percent of the net employment growth in the US economy" between 2005 and 2015 ("The Rise and Nature," 7). However, later on, the Bureau of Labor Statistics and Katz and Krueger ("The Rise and Nature") drew more moderate conclusions about the growth in these types of employment arrangements. However, these revised data have been contested along several lines. First, there are good reasons to believe that the updated numbers undercount and don't capture people who hold multiple jobs (Lenhart, "Is Gig Work"). Second, even if contingency didn't increase

in real numbers, perceptions of its rise still affect labor market decisions (Osterman, *Securing Prosperity*).

16. Cottom, "Nearly 6 Decades."
17. Shade, "What the Hustle Looks Like."
18. Dyson and Williams, *Jay-Z*, 17.
19. As Jackson ("Rethinking Repair," 223) explains, citing Star and Strauss ("Layers of Silence"), "Articulation supports the smooth interaction of parts within complex sociotechnical wholes, adjusting and calibrating each to each . . . sorting out ontologies on the fly rather than mixing and matching between fixed and stable entities. Articulation lives first and foremost in practice, not in representation."
20. Star and Straus, "Layers of Silence"; see also Gregg, *Work's Intimacy*, 167, on discounted labor practices.
21. Schwalbe et al., "Generic Processes," 426.
22. Braverman, *Labour and Monopoly Capitalism*, 92.
23. Snyder, *The Disrupted Workplace*.
24. On self-branding, see Vallas and Christin, "Work and Identity"; Gershon, *Down and Out*; and Lane, *A Company of One*.
25. Baym, "Connect with Your Audience!."
26. Gregg, *Work's Intimacy*, 168.
27. Gonzales, "Technology Maintenance," 2.
28. Marwick, *Status Update*, 190.
29. Madianou and Miller, *Migration and New Media*.
30. Foucault, *Discipline and Punish*.
31. Venkatesh, "Doin' the Hustle," 93.
32. A household manager position may oftentimes be combined with a nanny position but will also include managing and organizing other household staff, coordinating household maintenance and repairs, shopping, and other domestic tasks.
33. Sennett (*The Craftsman*, 9) observes that "craftsmanship cuts a far wider swath than skilled manual labor; it serves the computer programmer, the doctor, and the artist; parenting improves when it is practiced as a skilled craft, as does citizenship. In all these domains, craftsmanship focuses on object standards, on the thing in itself." While I share Sennett's capacious application of craft across many different types of work and labor, he has a timeless definition of craft, stating that the culture of late capitalism "militates against the ideal of craftsmanship, that is, learning to do just one thing really well; such commitment can often prove economically destructive" (Sennett, *The Craftsman*, 4). Sennet sees social and economic conditions as either barriers or facilitators of craft, while I understand the digital hustle as a craft practice that is indelibly marked by a particular set of cultural and economic conditions.
34. Hodson, *Dignity at Work*, 142.
35. Hodson, *Dignity at Work*, 142.
36. Thrift, *Knowing Capitalism*; Neff, *Venture Labor*; Duffy, *(Not) Getting Paid to Do*.
37. Sennett and Cobb, *The Hidden Injuries of Class*, 64, 154.
38. Lamont, *The Dignity of Working Men*; Sennett, *Culture of the New Capitalism*.

39. For example, for long-haul truckers, an ethic of "running hard" by taking infrequent breaks and driving fast to make shipping deadlines, is a source of pride (Snyder, *The Disrupted Workplace*). As Petriglieri, Ashford, and Wrzesniewski found, productivity itself, rather than belonging to a particular occupational or organizational group, was important to independent workers' ability to sustain a work identity. Petriglieri, Ashford, and Wrzesniewski, "Agony and Ecstasy."

40. Practices aren't the only things contingent workers use to form their identities (see Petriglieri, Ashford, and Wrzesniewski, "Agony and Ecstacy"; Barley and Kunda, *Gurus, Hired Guns*); however, this perspective has been missing in current explanations about the connection between digital technology and gig work. My understanding is closer to Venkatesh (*Off the Books*), who found that under-the-table workers drew on collective moral boundaries about acceptable and unacceptable ways to survive and thrive in communities of concentrated economic disadvantage. See Petriglieri, Ashford, and Wrzesniewski, "Agony and Ecstasy."

41. Vallas and Christin, "Work and Identity," 10.

42. Wenger, *Communities of Practice*.

Chapter 2

1. McDonald's and several other large chain restaurants offer free Wi-Fi Internet access, an increasingly valuable and scarce resource in low-income communities where low-wage workers live with their families. Troianovski, "Web-Deprived Study."

2. Sugie, "Work as Foraging."

3. Ling, *Taken for Grantedness*. On the use of digital technologies by precarious workers outside of the United States, see Donner, *After Access*; and Webster et al., *Grounding Globalization*.

4. Law and Peng, "Mobile Networks"; Qiu, *Working-Class Network Society*; Wallis, *Technomobility in China*.

5. Burrell, *Invisible Users*.

6. Greene, "Discovering the Divide."

7. Crain et al., "Conceptualizing Invisible Labor," 6.

8. MetroPCS is now owned by T-Mobile. In Washington, DC, I also recruited participants at a grocery store in this same neighborhood.

9. Lewis, *Handheld Device Ownership*.

10. Anderson, *Mobile Technology*.

11. See, Lewis, *Handheld Device Ownership*; Madden, *Privacy, Security*.

12. In 2012, Walmart was the largest retailer for prepaid phones. Fitchard, "Smartphones Make Big Gains."

13. As Tsetsi and Rains ("Smartphone Internet Access," 5) point out, "[a]lthough smartphones still may be relatively costly, they are typically less so than a laptop or desktop PC, especially because they are coupled with mobile Internet service. Laptop and desktop PCs, on the other hand, require the additional cost of a home broadband

connection for Internet access (or an obsolete dial-up modem). The lower financial barriers associated with smartphones, therefore, make them more accessible to disadvantaged groups than PCs requiring a broadband connection."

14. The National Digital Inclusion Alliance, and other researchers, have conducted research that has detailed these practices in Ohio and Texas, resulting in formal complaints being filed with the Federal Communications Commission. See Callahan, "AT&T's Digital Redlining"; and Tomer et al., *Digital Prosperity*.

15. See Martin, "Computer and Internet Use"; and Tomer and Fishbane, "Cleveland's Digital Divide."

16. I had several interviewees recount experiences of being confused and feeling misled around phone leasing or loan programs. Similar issues were also independently reported in Shahani, "What You Need to Know."

17. Stempel, "New York City Sues."

18. Ehrenreich, *Nickel and Dimed*.

19. On "predatory inclusion," see Seamster and Charron-Chénier, "Predatory Inclusion"; and Taylor, *Race for Profit*. On "reverse redlining," see Fisher, "Target Marketing."

20. To put these data limits in perspective, at the time of this writing, someone looking for a job using 2GB of data might be able to browse online job sites for four hours, spend two hours on Facebook, upload two to three resumes, and have two thirty-minute video interviews before using up their entire data allowance for one month. These estimates are based on typical data usage.

21. Between 2012 and 2013, prepaid phone subscriptions increased by 12 percent, while annual contracts stayed steady. Only half of prepaid customers had household incomes of more than $35,000 compared to 76 percent of annual contract customers. The Federal Communication Commission estimates that prepaid connections make up 20 percent of total connections in the United States. In 2012, Walmart was the largest retailer for prepaid phones. Tracfone and T-Mobile are the largest retailers of prepaid phones and services, together making up over half of the market. Lifsher, "More Cellphone Users"; CTIA, *Wireless Industry Overview*; Kovacs, *Competition in the US*; Federal Communications Commission, *FCC Fact Sheet*, 6.

22. Although prepaid phones have lower up-front fees, phone calls, text messages, and data usually cost more when compared with plans that require annual contracts, especially for those that use their phones regularly.

23. Seventy-five percent of the low-wage workers I interviewed were prepaid subscribers, and every one of them mentioned an instance of running out of data at an inconvenient time. Of the 25 percent of low-wage workers who were postpaid subscribers, 80 percent mentioned missing or being late on payments and having service cut off or facing the threat of it.

24. Napoli and Obar, "Emerging Mobile Internet Underclass," 330.

25. Federal Communications Commission, *20th Annual Report*, 39.

26. Cottom, *Lower Ed*.

27. Federal Trade Commission, "Debt Relief."

28. Gangadharan, "Dowside of Digital Inclusion."

29. Additionally, Zenith, a media buying agency, found that spending on mobile advertising grew by 35 percent in 2017. Anon., "Mobile Share of Advertising."
30. Gandy, *Panoptic Sort*, 1–2.
31. Couldry and Mejias, *Costs of Connection*; Federal Trade Commission, *Data Brokers.*
32. Madden et al., "Privacy, Poverty."
33. Forty-eight percent of people who depend on their smartphone for their primary Internet access report that they've lost access because of the "financial burden" of paying for service compared to 21 percent of people who have other ways of accessing the Internet. Fifty-one percent of smartphone-dependent users report losing service because they reached the maximum amount of data they were allowed to use on their plans compared to 35 percent with other means of accessing the Internet. Smith, *US Smartphone Use.*
34. Relying on secondhand technologies is a faster route to access, but it means that low-income populations are more likely to have outdated hardware, which only leads to more maintenance issues in the future. Gonzales, "The Contemporary US Digital Divide"; Horst and Miller, "From Kinship to Linkup".
35. In her comprehensive study of mobile phone use at work, Keri Stephens points out a mismatch between the widespread expectations among managers for their low-wage workers to be consistently connected and workers' ability to maintain their connectivity. Stephens, *Negotiating Control*, ch. 9.
36. Nationwide, 63 percent of rural residents have broadband Internet at home (compared to 73 percent of all American adults). Nationally, 81 percent of Americans own smartphones while only 71 percent of rural residents do, and my interviewees reflected this trend. Pew Research Center, *Demographics of Home Internet*; Pew Research Center, *Mobile technology and Home Broadband.* However, at the time of my interviews, smartphones were not nearly as commonplace among my interviewees in Mainville as they were in either the urban or suburban sites. Some owned smartphones, others had feature phones, and many cycled through predictable periods of having their service shut off and being able to turn it back on again the next week.
37. Mainville's unemployment rate is 8 percent (compared to 5 percent nationally), and it has a 31.7 percent poverty rate (compared to 14.8 percent nationally). For complete descriptive statistics on the field sites, see the methododological appendix, Figure Y.
38. Gonzales, "Contemporary US Digital Divide."
39. In 2016, 60.7 percent of children living below or close to the poverty line lived in homes without a landline. Blumberg and Luke, *Wireless Substitution.*
40. Standing, "Tertiary Time," 11.
41. In 2018, a confrontation in a Dunkin' Donuts in Virginia between a Black man using the store's Wi-Fi without a purchase and a store manager generated national news coverage. Subsequently, after two Black men who were waiting for a meeting at a Philadelphia Starbucks were arrested after refusing to leave the store, the company changed their store policies, which previously left decisions up to store managers, to allow people to sit and use the bathrooms without purchase. However, many new Starbucks stores are being built without customer-accessible outlets, and higher

traffic locations are covering them up or otherwise making them unavailable. Calfas, "Starbucks Is Closing"; Johnson, "Customers Say WeHo Starbucks"; Puhak, "Virginia Dunkin' Donuts."

42. Stewart, "Black Codes"; Goluboff, "Starbucks, LA Fitness."
43. Castells, *Rise of the Network Society*.
44. Castells, *Rise of the Network Society*.
45. Desmond, *Evicted*, 306.
46. Schradie, *Revolution That Wasn't*, 18; see also Rodino-Colocino, "Laboring under Digital Divide."

Chapter 3

1. Bourdieu, "Forms of Capital."
2. Bourdieu, "Forms of Capital."
3. As I explained in the last chapter, research on low-wage workers and digital technologies is usually done under the rubric of research about populations in the "global south," such as migrant workers, or in scholarly research about digital inequalities that does not focus on issues related to work but often focuses on lack of Internet access or digital skills for people who just so happen to be working.
4. Comparative research that examines similarities and differences between a well-studied and an understudied group, as this research does, runs the risk of reproducing the marginalization it seeks to understand. As Choo and Ferree point out, research that includes underrepresented groups to "give voice" to their experiences but neglects to investigate the construction of the "mainstream" risks reproducing notions of the dominant group as the standard. By examining the privileges that sustain the assumption of ubiquitous white-collar connectivity, this chapter uncovers the social contexts that produce this assumption. Choo and Ferree, "Practicing Intersectionality."
5. Feagin, *White Racial Frame*.
6. Patricia Hill Collins talks about this kind of analytical shift as being facilitated by moving the experiences of marginalized groups "from the margins to the center." While this chapter focuses on the experiences of a relatively powerful group of workers, it decenters their experiences by treating their circumstances as constructed instead of natural. Collins, *Black Feminist Thought*.
7. McIntosh, "White Privilege," 2.
8. Hargittai, "Second-Level Digital Divide"; Helsper, "Gendered Internet Use"; Mossberger et al., *Virtual Inequality*; Pearce and Rice, "Digital Divides."
9. Zillien and Hargittai, "Digital Distinction"; Helsper, "Corresponding Fields Model"; Selwyn, "Reconsidering Political."
10. In the case of precarious forms of gig work, the use of fields is especially helpful. As David Swartz explains, the concept of "fields" helped Bourdieu "cover social worlds where practices are only weakly institutionalized and boundaries are not well established." Swartz, *Culture and Power*, 120.

11. Chesley, "Blurring Boundaries?"; Towers et al., "Time Thieves"; Moen et al., "Time Work"; Perlow, *Sleeping with Your Smartphone*; Turkle, *Alone Together*.

12. Kalleberg, "Nonstandard Employment Relations."

13. See the methodological appendix for a more detailed discussion of these categories and descriptive statistics from participants in this research.

14. I began asking these questions as conversational segues between the interview and the postinterview survey after I turned my audio recorder off, but I found the patterns so compelling that I began to record the answers to understand this pattern. As a result, I only recorded this data for about 75 percent of my interviews. Nonetheless, the pattern was clear: 80 percent of high-wage workers said they weren't sure of the amount of their monthly bill or gave me a rough estimate and couldn't (or wouldn't) tell me the specific amount of their previous bills. While less than half reported being enrolled in auto-payments, every high-wage worker except one said they didn't check their bill monthly. This stands in stark contrast to the low-wage workers who, aside from five interviewees who had someone else who paid their bill, each reported the specific amount of their monthly bill and would more frequently discuss offerings from competing carriers or chat about switching carriers.

15. Bowker and Star, *Sorting Things Out*, 299.

16. Khan, *Privilege*. As Bourdieu poetically puts it, "Innocence is the privilege of those who move in their field of activity like a fish in water" Bourdieu, "Forms of Capital," 28.

17. Barley and Kunda, *Gurus, Hired Guns, and Warm Bodies*; Popeil, " 'Boundaryless' in the Creative Economy."

18. Khan writes that ease naturalizes privilege because of an irony at the heart of it: "ease requires hard, systematic work, yet the result should be 'natural' and effortless." Khan, *Privilege*, 112–113.

19. Interestingly, while the entrepreneurs that Marwick followed for her book on social media in 2010 were explicit about self-branding, the influencers and micro-celebrities that Brooke Duffy interviewed in 2015 had a much more conflicted relationship to the idea of themselves as a "brand," as the authenticity mandate of social media platforms has only deepened. See Duffy, *(Not) Getting Paid*; Marwick, *Status Update*.

20. Granovetter, "Strength of Weak Ties"; Putnam, *Bowling Alone*.

21. On early scholarly concerns about technology's negative influence on social capital, see Putnam, *Bowling Alone*. For later scholarship on the positive influence of the Internet on social capital, see Ellison et al., "Benefits of Facebook"; Katz and Rice, *Social Consequences*; and Wellman et al., "Capitalizing on the Internet."

22. Hargittai, "Second-Level Digital Divide."

23. Hargittai and Hinnant, "Digital Inequality."

24. As Bourdieu explains, there are some fields that provide both the "rules of the game" and the social space that decides which cards players are dealt in the first place. Bourdieu, *Outline of a Theory*.

25. Mesch, "Minority Status."

26. Certeau, *Practice of Everyday Life*, 25.

27. Rafalow, "Disciplining Play," 1447.

28. Florida, *Rise Of the Creative Class*; Drucker, *Post-Capitalist Society*.

29. Greene, "Discovering the Divide."

Chapter 4

1. Hochschild, *The Managed Heart.*
2. Rae's strategy shared some similarities with what Melissa Gregg calls "presence bleed" among white-collar professionals. She found that, by using their laptops and smartphones at the kitchen table or during family meals, these workers' performances as professionals "seep[ed] in to and coexist[ed] alongside other spheres of life." Gregg, *Work's Intimacy*, 3.
3. These two forms of control are sometimes referred to as rational and normative control. Barley, "Design and Devotion"; Kunda, *Engineering Culture.*
4. Edwards called this "technical control." Edwards, *Contested Terrain.*
5. Gouldner, *Patterns of Industrial Bureaucracy*; Hobsbawm, "The Machine Breakers"; Roy, "Quota Restriction."
6. Gregg, *Work's Intimacy*; Barley et al., "E-Mail as a Source"; Qiu, "The Wireless Leash"; Wallis, *Technomobility in China.*
7. O'Reilly and Chatman, "Culture as Social Control"; Kunda, *Engineering Culture.*
8. Scott, *Weapons of the Weak.* On "micro-resistance," see also Fleming and Spicer, *Contesting the Corporation.*
9. Scott, *Weapons of the Weak*, 30.
10. In his foundational work on precarious workers, Standing (2011) observes that precarious workers are anomic and lack collective, stable work-based identities. Standing, *The Preacariat.*
11. Scott pointed out that forms of individualized resistance maintain "symbolic compliance" with the ruling norms while "skirt[ing] the edge of impropriety." Scott, *Weapons of the Weak*, 26. See also Ticona, "Strategies of Control."
12. While Rae and Kitty were both service workers, I also saw these practices among those low-wage workers in administrative "pink-collar" work. Many refrained from using their work-provided computers for fear of surveillance but would still use their phones to resist managerial expectations of full attentiveness. Higher wage white-collar workers also admitted getting "distracted" with social media and games at work but most often not with the aim of resistance. For example, after telling me that she got nervous about checking superficial gossip sites at work because her boss has the ability to "audit" her Internet history, Penelope, a corporate lawyer in Washington, DC, defended this practice because she needed to "take breaks," saying, "I would say when I find myself either drifting mentally or because there is a lot of repetition and monotony, if I find that I'm not looking at a document and I'm just assuming that I know what's in it, that would be a good time to take check Gawker or something because that leads to mistakes." While other high-wage workers like Penelope were engaged in many of the same practices as the low-wage workers, their reasons had more to do with improving their ability to do their jobs, not resisting control.
13. Pollert, *Girls, Wives, Factory Lives.*
14. A "handle" is a unique username that is associated with individual Twitter users.
15. Senft, "Microcelebrity."
16. Marwick and boyd, "I Tweet Honestly."
17. Jenkins, *Convergence Culture.*

18. BlackPlanet is a Black social networking site launched in 2001 that had discussion forums for political issues, dating, and jobs. Byrne, "Public Discourse, Community Concerns."

19. Kunda, *Engineering Culture*.

20. Lane, *Company of One*; Petriglieri et al., "Agony and Ecstasy."

21. Foucault, *Birth of Biopolitics*, 226.

22. Mazmanian et al., "Work Without End?"

23. Barley and Kunda. *Gurus, Hired Guns, and Warm Bodies*, 238–241.

24. Gregg, *Counterproductive*, 6. See also Sharma, *In the Meantime*; Wajcman, *Pressed for Time*.

25. Nippert-Eng, *Islands of Privacy*; Zerubavel, "Private Time."

26. Nippert-Eng, *Islands of Privacy*; Wajcman, *Pressed for Time*.

27. Derber, *The Pursuit of Attention*. James, a low-wage hustler in Washington, DC, encapsulated this dynamic when I asked him if he thought of himself as being "on call" to his many clients, saying: "I'm on call like a CEO, you know? I'm running the show, so I have to have the phone on me . . . except [laughing] I don't have that CEO money yet. I have to text back quick or else. Once I get there though, [laughs] I'm *never* answering my phone! When you have money, you don't have to answer, but me? For now, I have to." James's quote points to the ways high-wage workers' status allowed them to disconnect while simultaneously reinforcing their status.

28. Hays, *Cultural Contradictions of Motherhood*.

29. This practice is similar to what researchers call practices of "connected presence" or "perpetual contact" that occur when physically separated communicative partners "gain presence" through the use of "mediated communication gestures on both sides." The technical affordances of mobile information and communications technologies, especially their small size and Internet connection, allow for partners to bend the constraints of space and cultivate intimacy in near real time through affective exchanges of text, pictures, and voice communications. On gender and kinkeeping, see Rosenthal, "Kinkeeping." On "connected presence," see Licoppe, "'Connected' Presence." On mobile phones and intimacy, see Katz and Aakhus, *Perpetual Contact*; and Schroeder, "Being There Together."

30. Hochschild and Machung, *The Second Shift*; Hochschild, *The Time Bind*.

31. Blair-Loy, *Competing Devotions*.

32. Crowe and Middleton, "Women, Smartphones and the Workplace."

33. Sirrom wasn't alone among the women I interviewed in blaming herself for failing to meet everyone's expectations. For the women in my study who were high-wage, white-collar gig workers, three-quarters blamed themselves in their discussions about tensions that emerged between expectations of tethered carework and expectations of being fully invested and distraction-free at work. Twelve of the sixteen high-wage women I spoke with mentioned blaming themselves when workplace expectations clashed with other kinds of cultural expectations about performing care with their technologies.

34. Gray, Suri, Ali, and Kulkarni, "The Crowd Is a Collaborative Network"; Irani and Silberman, "Turkopticon: Interrupting Worker Invisibility in Amazon Mechanical

Turk"; Qiu, "Social Media on the Picket Line"; Schradie, *The Revolution That Wasn't*; Wood, Lehdonvirta, and Graham, "Workers of the Internet Unite?".

35. Scott, *Weapons of the Weak*, 27.

36. Portwood-Stacer, "Media Refusal and Conspicuous Non-Consumption."

37. Mazmanian, Orlikowski, and Yates, "The Autonomy Paradox: The Implications of Mobile Email Devices for Knowledge Professionals;" Perlow, *Sleeping with Your Smartphone*.

38. Browne, *Dark Matters*, 10.

39. Collins, *Black Feminist Thought*; Crenshaw, "Mapping the Margins."

Conclusion

1. Morris, "Super Bowl Ad Cost 2021."

2. Slaughter, "The Gig Economy"; Sundararajan, *The Sharing Economy*.

3. Ravenelle, *Hustle and Gig*; Scholz, *Uberworked and Underpaid*; Schor, *After the Gig*.

4. van Doorn, "Platform Labor"; Irani, "Cultural Work of Microwork"; Rosenblat, *Uberland*; Schor, *After the Gig*.

5. Hyman, *Temp*.

6. Greene and Ajunwa, "Platforms at Work"; Levy and Barocas, "Refractive Surveillance"; Stanley, "ACLU White Paper"; Van Oort, "Emotional Labor of Surveillance."

7. Christin, "Counting Clicks"; Katz and Gonzalez, "Toward Meaningful Connectivity"; Lane, *The Digital Street*; Rafalow, *Digital Divisions*; Greene, *The Promise of Access*.

8. Dimaggio et al., "From Unequal Access"; Hargittai, "Second-Level Digital Divide"; Pearce and Rice, "Digital Divides"; van Dijk, "Digital Divide Research." For a critique of the individual bent of studying exclusion in digital inequalities research, see Helsper, "Social Relativity."

9. Dunbar-Hester, *Hacking Diversity*, Hoffmann, "Terms of Inclusion."

10. Benjamin, *Race After Technology*; Hoffman, "Terms of Inclusion."

11. Even in understanding what appear to be clear cut issues of "first level" digital divides in access, it's still crucial that we examine the role of social institutions that produce racialized inequalities, such as unequal access to credit, or unequal policing. See, Yang, Ticona, Lelkes, "Policing the Digital Divide."

12. Kalleberg, *Good Jobs, Bad Jobs*; Sennett, *Culture of the New Capitalism*.

13. Louise Seamster describes "predatory inclusion" as the inclusion of formerly excluded populations, "on terms that negate the advantages of incorporation." See Seamster, "Black Debt, White Debt." See also Gangadharan, "Downside of Digital Inclusion"; and Fourcade and Healy, "Classification Situations."

14. On the possibilities of data being used for liberatory inclusion in the interests of workers, see Miller and Adler-Bell, *Datafication of Employment*.

15. Accominotti, Storer, and Khan observe that, in the Gilded Age, when elite institutions, like the New York Philharmonic, opened up to outsiders and (somewhat) democratized access to cultural experiences previously reserved for the ultra-rich, they saw

what that new and old elites didn't mingle in the concert hall, but that newcomers were included on segregated terms. Accominotti et al., "How Cultural Capital Emerged."

16. For important exceptions, see Rafalow, *Digital Divisions*; Robinson, "A Taste for the Necessary"; and Schradie, *Revolution That Wasn't*.

17. In this quote, Cottom is summarizing contributions of Black feminists, such as Patricia Hill Collins, to her approach to investigating digital technologies and power, an approach that centers the experiences of marginalized groups. See Cottom, "Black CyberFeminism," 212.

18. As Ellen Helsper points out, digital inequalities need to be understood as relative to particular social contexts. Helsper, "Social Relativity of Digital Exclusion."

19. McPherson, "Digital Humanities So White?"

20. See Hui et al., "Community Collectives."

Methodological Appendix

1. Wiatrowski, "BLS Measures."

2. Kalleberg, "Precarious Work," 2.

3. Bureau of Labor Statistics, "FAQs about Data."

4. The distinction between a participant's sense of their main type of work and their main source of income was complex. For example, a participant in New York who had graduated from fine arts school and was writing a book of poetry considered writing her main type of work because it's what she had been trained to do and was most proud of, but her main sources of income were as a home organizer and other odd jobs she got on several different online labor platforms. I would ask, "What do you do to earn money?" and if they answered by describing one job, I would probe for others and often explicitly used the terms "side hustles" and "side jobs."

5. Kopf, "Slowly but Surely."

6. FCC, "2018 Broadband Deployment Report."

7. Ravenelle, *Hustle and Gig.*

8. In my rural fieldsite, I encountered a healthy suspicion of myself and toward academic research in general when I approached strangers in stores and at community events. When possible, I took along a local informant, who would sometimes introduce me to acquaintances of theirs who came through the stores or that we ran into at community events.

9. See Jeff Lane's discussion of the important, but economically dubious, role of these kinds of phone retailers for teens and people without credit in Harlem, New York. Lane, *The Digital Street*, 22–24.

10. The Internal Revenue Service evaluates the control of work along twenty different criteria to determine independent contractor status. Kalleberg and Dunn suggest that autonomy in the execution of work, scheduling, financial transactions, and duration are important to understanding the quality of gig work. Independent Revenue Service, "Independent Contractor Defined"; Kalleberg and Dunn, "Good Jobs, Bad Jobs."

11. Ticona, "Strategies of Control."
12. These phrases are paraphrases from my fieldnotes, not actual quotes, as at this point in recruitment, interviewees had not consented to participate in the study.
13. Spradley, *The Ethnographic Interview*.
14. Three of my interviews were conducted in small groups of interviewees who were either close friends or family members (a total of eight participants). Four individual interviews were conducted over Skype.
15. Charmaz, *Constructing Grounded Theory*.
16. Pugh, *Longing and Belonging*, 44.
17. Jerolmack and Khan, "Talk Is Cheap."
18. Pugh, "What Good Are Interviews?," 50.

Bibliography

Accominotti, Fabien, Shamus R. Khan, and Adam Storer. "How Cultural Capital Emerged in Gilded Age America: Musical Purification and Cross-Class Inclusion at the New York Philharmonic." *American Journal of Sociology* 123, no. 6 (2018): 1743–1783.

Adler Patricia, and Adler Peter. "Shifts and Oscillation in Deviant Careers: The Case of Upper-Level Dealers and Smugglers." *Social Problems* 31 (1983): 195–207.

Anderson, Monica. *Mobile Technology and Home Broadband 2019*. Washington, DC: Pew Research Center, 2019.

Anderson, Monica. *Technology Device Ownership: 2015* Washington, DC: Pew Research Center, 2015.

Anon. "Lifeline Program Statistics." *Universal Service Administrative Co.*, 2019.

Anon. "Managing the Future of Work." Harvard Business School. Podcast. 2019. https://www.hbs.edu/managing-the-future-of-work/podcast/Pages/podcast-details.aspx?episode=11567690.

Anon. "Obama Donors Get Obama Phone Kickbacks." *The Rush Limbaugh Show*, October 8, 2012.

Anon. "What We're Up Against: The Obama Phone." *The Rush Limbaugh Show*, June 22, 2012.

Arcy, J. "Emotion Work: Considering Gender in Digital Labor." *Feminist Media Studies* 16 (2016): 365–368.

Avery, D., and Crain M. "Branded: Corporate Image, Sexual Stereotyping, and the New Face of Capitalism." *Duke Journal of Gender Law & Policy* 14 (2007): 13–123.

Barley, Stephen, and Gideon Kunda. *Gurus, Hired Guns, and Warm Bodies: Itinerant Experts in a Knowledge Economy*. Princeton, NJ: Princeton University Press, 2004.

Barley, Stephen R., and Gideon Kunda. "Design and Devotion: Surges of Rational and Normative Ideologies of Control in Managerial Discourse." *Administrative Science Quarterly* 37, no. 3 (1992): 363–399. doi: 10.2307/2393449.

Barley, Stephen, Debra Meyerson, and Stine Grodal. "E-Mail as a Source and Symbol of Stress." *Organization Science* 22, no. 4 (2011): 887–906.

Baym, Nancy K. "Connect With Your Audience! The Relational Labor of Connection." *The Communication Review* 18 (2015): 14–22.

Beck, Ulrich, and Elisabeth Beck-Gernsheim. *Individualization*. London: Sage, 2002.

Becker, Howard. "Arts and Crafts." *American Journal of Sociology* 83, no. 4 (1978): 862–889.

Bell, Daniel. *The Coming of Post-Industrial Society: A Venture in Social Forecasting*. New York: Basic Books, 1976.

Benjamin, Ruha. *Race After Technology: Abolitionist Tools for the New Jim Code*. Medford, MA: Polity, 2019.

Blair-Loy, Mary. "Work Without End?: Scheduling Flexibility and Work-to-Family Conflict Among Stockbrokers." *Work and Occupations* 36, no. 4 (2009): 279–317.

Blair-Loy, Mary. *Competing Devotions: Career and Family among Women Executives*. Cambridge, MA: Harvard University Press, 2003.

Bloor, David. *Knowledge and Social Imagery.* Chicago: University of Chicago Press, 1976.

Blumberg, Stephen, and Julian Luke. *Wireless Substitution: Early Release of Estimates from the National Health Interview Survey, July—December 2016.* Washington, DC: National Center of Health Statistics, 2016.

Bonilla-Silva, Eduardo. "Rethinking Racism: Toward a Structural Interpretation." *American Sociological Review* 62, no. 3 (1997): 465–480.

Bourdieu, Pierre. "The Forms of Capital." In *Handbook of Theory and Research for the Sociology of Education,* edited by John G. Richardson, 241–258. Westport, CT: Greenwood Press, 1986.

Bourdieu, Pierre. *Distinction: A Social Critique of the Judgment of Taste.* Translated by Richard Nice. Cambridge, MA: Harvard University Press, 1984.

Bourdieu, Pierre. *Outline of a Theory of Practice.* Translated by Richard Nice. Cambridge, UK: Cambridge University Press, 1977.

Bowker, Geoffrey C., and Susan Leigh Star. *Sorting Things Out: Classification and Its Consequences.* Cambridge, MA: MIT Press, 2000.

Braverman, Harry. *Labor and Monopoly Capital: The Degradation of Work in the Twentieth Century.* New York: Monthly Review Press, 1998.

Browne, Simone. *Dark Matters: On the Surveillance of Blackness.* Durham, NC: Duke University Press, 2015.

Burawoy, Michael. *Manufacturing Consent: Changes in the Labor Process Under Monopoly Capitalism.* Chicago: University of Chicago Press, 1982.

Bureau of Labor Statistics. "FAQs about Data on Contingent and Alternative Employment Arrangements (CPS)." *Bureau of Labor Statistics.* https://www.bls.gov/cps/contingent-and-alternative-arrangements-faqs.htm#contingent.

Burrell, Jenna. *Invisible Users: Youth in the Internet Cafés of Urban Ghana.* Cambridge, MA: MIT Press, 2012.

Byrne, Dara N. 2007. "Public Discourse, Community Concerns, and Civic Engagement: Exploring Black Social Networking Traditions on BlackPlanet.Com." *Journal of Computer-Mediated Communication* 13, no. 1 (2007): 319–340.

Calfas, Jennifer. "Starbucks Is Closing All Its US Stores for Diversity Training Day. Experts Say That's Not Enough." *TIME Magazine,* May 28, 2018. https://time.com/5287082/corporate-diversity-training-starbucks-results/.

Callahan, Bill. "AT&T's Digital Redlining of Cleveland." *National Digital Inclusion Alliance,* 2017. https://www.digitalinclusion.org/blog/2017/03/10/atts-digital-redlining-of-cleveland/.

Castells, Manuel. *The Internet Galaxy: Reflections on the Internet, Business, and Society.* Oxford: Oxford University Press, 2001.

Castells, Manuel. *The Rise of the Network Society.* New York: Blackwell, 1996.

Charmaz, Kathy. *Constructing Grounded Theory: A Practical Guide through Qualitative Analysis.* Thousand Oaks, CA: Sage, 2006.

Chesley, Noelle. "Blurring Boundaries? Linking Technology Use, Spillover, Individual Distress, and Family Satisfaction." *Journal of Marriage and Family* 67 (2005): 1237–1248.

Chesley, Noelle. "Families in a High-Tech Age: Technology Useage Patterns, Work and Family Correlates, and Gender." *Journal of Family Issues* 27 (2006): 587–608.

Choo, Hae Yeon, and Myra Marx Ferree. "Practicing Intersectionality in Sociological Research: A Critical Analysis of Inclusions, Interactions, and Institutions in the Study of Inequalities." *Sociological Theory* 28, no. 2 (2010): 129–149. https://doi.org/10.1111/j.1467-9558.2010.01370.x.

Christin, Angèle. "Counting Clicks: Quantification and Variation in Web Journalism in the United States and France." *American Journal of Sociology* 123, no. 5 (2018): 1382–1415.

Chun, Wendy Hui Kyong. *Updating to Remain the Same: Habitual New Media.* Cambridge, MA: MIT Press, 2016.

Cloward, Richard. "Illegitimate Means, Anomie, and Deviant Behavior." *American Sociological Review* 24 (1959): 164–176.

Collins, Patricia Hill. *Black Feminist Thought: Knowledge, Consciousness, and the Politics of Empowerment.* New York: Routledge, 1990.

Cooper, Marianne. *Cut Adrift: Families in Insecure Times* Oakland: University of California Press, 2014.

Cottom, Tressie McMillan. "Nearly 6 Decades After the Civil Rights Act, Why Do Black Workers Still Have to Hustle to Get Ahead?" *Time,* February 20, 2020. https://time.com/5783869/gig-economy-inequality/

Cottom, Tressie McMillan. "Black CyberFeminism: Intersectionality, Institutions, and Digital Sociology." In *Digital Sociologies,* edited by Jessie Daniels, Karen Gregory, Tressie McMillan Cottom, 211–232. Bristol, UK: Policy Press, 2017.

Cottom, Tressie McMillan. *Lower Ed: The Troubling Rise of For-Profit Colleges in the New Economy.* New York: The New Press, 2017.

Couldry, Nick, and Ulises A. Mejias. *The Costs of Connection: How Data Is Colonizing Human Life and Appropriating It for Capitalism.* Stanford, CA: Stanford University Press, 2019.

Crain, Marion, Winifred Poster, and Miriam Cherry. "Conceptualizing Invisible Labor." In *Invisible Labor: Hidden Work in the Contemporary World,* edited by Marion Crain, Winifred Poster, and Miriam Cherry, 3–27. Oakland: University of California Press, 2016.

Crenshaw, Kimberlé. "Mapping the Margins: Intersectionality, Identity Politics, and Violence against Women of Color." *Stanford Law Review* 43 (1990): 1241–1300.

Crowe, Rachel, and Catherine Middleton. "Women, Smartphones and the Workplace." *Feminist Media Studies* 12, no.4 (2012): 560–569.

Cellular Telecommunications and Internet Association. *Wireless Industry Overview.* Washington, DC: The Wireless Association, 2012.

de Certeau, Michel. *The Practice of Everyday Life.* Translated by Steven Rendall. Berkeley: University of California Press, 1984.

Derber, Charles. *The Pursuit of Attention: Power and Ego in Everyday Life.* New York: Oxford University Press, 1979.

Desmond, Matthew. *Evicted: Poverty and Profit in the American City.* New York: Crown, 2016.

Dimaggio, Paul, Eszter Hargittai, Coral Celeste, and Steven Shafer. "From Unequal Access to Differentiated Use: From Unequal Access to Differentiated Use." In *Social Inequalities,* edited by Kathryn Neckerman, 355–400. New York: Russell Sage Foundation, 2004.

Donner, Jonathan. *After Access: Inclusion, Development, and a More Mobile Internet.* Cambridge, MA: MIT Press, 2015.

Drucker, Peter F. *Post-Capitalist Society.* New York: HarperBusiness, 1994.

Drucker, Peter F. *Landmarks of Tomorrow: A Report on the New.* New York: Harper and Row, 1959.

Duffy, Brooke E. *(Not) Getting Paid to Do What You Love: Gender, Social Media, and Aspirational Work.* New Haven, CT: Yale University Press, 2017.

Dunbar-Hester, Christina. *Hacking Diversity: The Politics of Inclusion in Open Technology Cultures.* Princeton, NJ: Princeton University Press, 2020.

Duxbury, Linda, Ian Towers, Christopher Higgens, et al. "From 9 to 5 to 24/7: How Technology Has Redefined the Workday." In *Information Resources Management: Global Challenges*, edited by Wai K. Law, 1–28. Hershey, PA: IGI Global, 2006.

Dyson, Michael Eric, and Williams Pharrell. *JAY-Z: Made in America*. New York: St. Martin's Press, 2019.

Edin, Kathryn, and Timothy J. Nelson. *Doing the Best I Can: Fatherhood in the Inner City*. Berkeley: University of California Press, 2013.

Edin, Kathryn, and Luke H. Shaefer. *$2.00 a Day: Living on Almost Nothing in America*. Reprint ed. New York: Mariner Books, 2016.

Edin, Kathryn, Laura Lein, and Christopher Jencks. *Making Ends Meet: How Single Mothers Survive Welfare and Low-Wage Work*. New York: Russell Sage Foundation, 1997.

Edwards, Richards. *Contested Terrain*. New York: Basic Books, 1979.

Ehrenreich, Barbara. *Nickel and Dimed: On (Not) Getting By in America*. New York: Picador, 2001.

Ellison, Nicole B., Charles Steinfield, and Cliff Lampe. "The Benefits of Facebook 'Friends:' Social Capital and College Students' Use of Online Social Network Sites." *Journal of Computer-Mediated Communication* 12, no. 4 (2007): 1143–1168.

Espeland, Wendy N., and Michael Sauder. "Rankings and Reactivity: How Public Measures Recreate Social Worlds." *American Journal of Sociology* 113 (2007): 1–40.

Eubanks, Virginia. *Digital Dead End: Fighting for Social Justice in the Information Age*. Reprint ed. Cambridge, MA: The MIT Press, 2012.

Farrell, Diana, Fiona Greig, and Amar Hamoudi. "The Evolution of the Online Platform Economy: Evidence from Five Years of Banking Data." *AEA Papers and Proceedings* 109 (2019): 362–366.

Feagin, Joe R. *The White Racial Frame: Centuries of Racial Framing and Counter-Framing*. New York: Routledge, 2013.

Federal Communications Commission. *20th Annual Report and Analysis of Competitive Market Conditions with Respect to Mobile Wireless, Including Commercial Mobile Services* Washington, DC: Government Printing Office, 2017.

Federal Communications Commission. *FCC Fact Sheet: Communications Marketplace Report*. Washington, DC: Government Printing Office, 2018.

Federal Communications Commission. *2018 Broadband Deployment Report*. Washington, DC: Government Printing Office, 2018.

Federal Trade Commission. *Data Brokers: A Call for Transparency and Accountability*. Washington, DC: Government Printing Office, 2014.

Federal Trade Commission. "Debt Relief and Credit Repair Scams." *Federal Trade Commission*. https://www.ftc.gov/news-events/media-resources/consumer-finance/debt-relief-credit-repair-scams.

Fisher, Linda. "Target Marketing of Subprime Loans: Racialized Consumer Fraud and Reverse Redlining." *Brooklyn Journal of Law & Policy* 18, no. 1 (2009): 121–155.

Fitchard, Kevin. "Smartphones Make Big Gains in Prepaid." *GIGAOM*, March 28, 2012. https://gigaom.com/2012/03/28/smartphones-make-big-gains-in-prepaid/.

Fleming, Peter, and André Spicer. *Contesting the Corporation: Struggle, Power and Resistance in Organizations*. Cambridge, MA: Cambridge University Press, 2007.

Florida, Richard. *The Rise of the Creative Class and How It's Transforming Work, Leisure, Community and Everyday Life*. Princeton, NJ: Basic Books, 2002.

Foucault, Michel. *The Birth of Biopolitics: Lectures at the Collège de France, 1978–1979*. Edited by Michel Senellart. Translated by Graham Burchell. New York: Picador, 2010.

Foucault, Michel. *Discipline & Punish: The Birth of the Prison.* New York: Vintage Books, 1995.

Fourcade, Marion, and Kieran Healy. "Classification Situations: Life-Chances in the Neoliberal Era." *Accounting, Organizations and Society* 38 (2013): 559–572.

Gandy, Oscar. 1993. *The Panoptic Sort: Political Economy of Personal Information.* Nashville, TN: Westview Press, 1993.

Gangadharan, Seeta Peña. "The Downside of Digital Inclusion: Expectations and Experiences of Privacy and Surveillance among Marginal Internet Users." *New Media & Society* 19, no. 4 (2015): 597–615. doi: 1461444815614053.

Gershon, Ilana. *Down and Out in the New Economy: How People Find (or Don't Find) Work Today.* Chicago: University of Chicago Press, 2017.

Gilliom, John. *Overseers of the Poor: Surveillance, Resistance, and the Limits of Privacy.* Chicago: University of Chicago Press, 2001.

Goluboff, Risa. "Starbucks, LA Fitness and the Long, Racist History of America's Loitering Laws." *Washington Post,* April 20, 2018. https://www.washingtonpost.com/news/post-nation/wp/2018/04/20/starbucks-la-fitness-and-the-racist-history-of-trespassing-laws/.

Gonzales, Amy. "The Contemporary US Digital Divide: From Initial Access to Technology Maintenance." *Information, Communication & Society* 19 (2016): 234–248.

Gonzales, Amy. "Technology Maintenance: A New Frame for Studying Poverty and Marginalization." In Proceedings of *Computer Human Interaction,* 289–294, *New York: Association for Computing Machinery, 2017.*

Gonzales A., L. Ems, and V. R. Suri. "Cell Phone Disconnection Disrupts Access to Healthcare and Health Resources: A Technology Maintenance Perspective." *New Media & Society* 18 (2016): 1422–1438.

Gonzales, Amy. "The Contemporary US Digital Divide: From Initial Access to Technology Maintenance." *Information, Communication & Society* 19, no. 2 (2016): 234–248.

Gouldner, Alvin W. *Patterns of Industrial Bureaucracy.* New York: Free Press, 1954.

Government Accountability Office. *Contingent Workforce: Size, Characteristics, Earnings, and Benefits.* Washington, DC: Government Accountability Office, April 20, 2015. https://www.gao.gov/products/gao-15-168r.

Granovetter, Mark. "The Strength of Weak Ties." *American Journal of Sociology* 78, no. 6 (1973): 1360–1380.

Gray, Mary L., Siddharth Suri, Syed Shoaib Ali, and Deepti Kulkarni. "The Crowd Is a Collaborative Network." In *Proceedings of the 19th ACM Conference on Computer-Supported Cooperative Work and Social Computing, CSCW '16,* 134–147. New York: Association for Computing Machinery, 2016.

Grazian, David. *On the Make: The Hustle of Urban Nightlife.* Chicago: University of Chicago Press, 2008.

Greene, Daniel. *The Promise of Access: Technology, Inequality, and the Political Economy of Hope.* Cambridge, MA: The MIT Press, 2021.

Greene, Daniel. "Discovering the Divide: Technology and Poverty in the New Economy." *International Journal of Communication* 10 (2016): 1212–1231.

Greene, Daniel, and Ifeoma Ajunwa. "Platforms at Work: Automated Hiring Platforms and Other New Intermediaries in the Organization of Work | Emerald Insight| Emerald Insight." *Research in the Sociology of Work* 33 (2019): 61–91.

Gregg, Melissa. *Counterproductive: Time Management in the Knowledge Economy.* Durham, NC: Duke University Press, 2018.

Gregg, Melissa. *Work's Intimacy*. Cambridge: Polity, 2011.

Hacker, Jacob. *The Great Risk Shift: The New Economic Insecurity and the Decline of the American Dream*. New York: Oxford University Press, 2008.

Hargittai, Eszter. "Digital Na(t)ives? Variation in Internet Skills and Uses among Members of the 'Net Generation.'" *Sociological Inquiry* 80 (2010): 92–113.

Hargittai, Eszter. "Second-Level Digital Divide: Differences in People's Online Skills." *First Monday* 7 (2002).

Hargittai, Eszter, and Amanda Hinnant. "Digital Inequality: Differences in Young Adults' Use of the Internet." *Communication Research* 35, no. 5 (2008): 602–621.

Hays, Sharon. *The Cultural Contradictions of Motherhood*. New Haven, CT: Yale University Press, 1996.

Helsper, Ellen Johanna. "The Social Relativity of Digital Exclusion: Applying Relative Deprivation Theory to Digital Inequalities." *Communication Theory* 27, no. 3 (2017): 223–242.

Helsper, Ellen Johanna. "A Corresponding Fields Model for the Links between Social and Digital Exclusion." *Communication Theory* 22, no. 4 (2012): 403–426.

Helsper, Ellen Johanna. "Gendered Internet Use Across Generations and Life Stages." *Communication Research* 37, no. 3 (2010): 352–374.

Hobsbawm, Eric. *Labouring Men: Studies in the History of Labour*. New York: Basic Books, 1965.

Hochschild, Arlie Russell. *Strangers in Their Own Land: Anger and Mourning on the American Right*. New York: The New Press, 2016.

Hochschild, Arlie Russell. *The Time Bind: When Work Becomes Home and Home Becomes Work*. New York: Holt Paperbacks, 2001.

Hochschild, Arlie Russell. *The Managed Heart*. Berkeley: University of California Press, 1983.

Hochschild, Arlie Russell, and Anne Machung. *The Second Shift*. New York: Penguin Books, 1989.

Hodson, Randy. *Dignity at Work*. Cambridge, UK: Cambridge University Press, 2001.

Hoffmann, Anna Lauren. "Terms of Inclusion: Data, Discourse, Violence." *New Media & Society* 2020. doi:1461444820958725.

Hollister, Matissa. "Employment Stability in the U.S. Labor Market: Rhetoric versus Reality." *Annual Review of Sociology* 37 (2011): 305–324.

Horst, Heather, and Daniel Millerl. "From Kinship to Link-up: Cell Phones and Social Networking in Jamaica." *Current Anthropology* 46 (2005): 755–778.

Horowitz, Juliana, Ruth Igielnik, and Tanya Arditi. *Trends in US Income and Wealth Inequality*. Washington, DC: Pew Research Center, 2020.

Hui, Julie, Barber Nefer Ra, Casey Wendy, et al. "Community Collectives: Low-tech Social Support for Digital-Engaged Entrepreneurship." In *Proceedings of ACM Human-Computer Interaction*, 1–15. New York: Association for Computing Machinery, 2020.

Hyman, Louis. 2018. *Temp: How American Work, American Business, and the American Dream Became Temporary*. New York: Viking, 2018.

Illouz, Eva. *Consuming the Romantic Utopia: Love and the Cultural Contradictions of Capitalism*. Berkeley: University of California Press, 1997.

Internal Revenue Service. "Independent Contractor Defined." *Internal Revenue Service*. https://www.irs.gov/businesses/small-businesses-self-employed/independent-contractor-defined.

Irani, Lilly. "The Cultural Work of Microwork." *New Media & Society* 17, no. 5 (2015): 720–739.

Irani, Lilly. "Difference and Dependence among Digital Workers: The Case of Amazon Mechanical Turk." *South Atlantic Quarterly* 114 (2015): 225–234.

Irani, Lilly C., and M. Six Silberman. "Turkopticon: Interrupting Worker Invisibility in Amazon Mechanical Turk." In *Proceedings of the SIGCHI Conference on Human Factors in Computing Systems*, 611–620. New York: Association for Computing Machinery, 2013.

Jackson, Steven. "Rethinking Repair." In *Media Technologies: Essays on Communication, Materiality, and Society*, 221–239. Cambridge, MA: MIT Press, 2014.

Jarrett, Kylie. *Feminism, Labour and Digital Media: The Digital Housewife.* New York: Routledge, 2015.

Jenkins, Henry. *Convergence Culture: Where Old and New Media Collide.* New York: New York University Press, 2008.

Jenkins, Henry. "Cyberspace and Race." *MIT Technology Review*, April 1, 2002. https://www.technologyreview.com/s/401404/cyberspace-and-race/.

Jerolmack, Colin, and Shamus Khan. 2014. "Talk Is Cheap: Ethnography and the Attitudinal Fallacy." *Sociological Methods & Research* 43, no. 2 (2014): 178–209.

Johnson, Matt. "Customers Say WeHo Starbucks Covered Power Outlets to Discourage Loitering." *KTTV*, March 17, 2017. http://www.foxla.com/news/local-news/custom ers-say-weho-starbucks-covered-power-outlets-to-discourage-loitering.

Kalleberg, Arne L. *Good Jobs, Bad Jobs: The Rise of Polarized and Precarious Employment Systems in the United States 1970s to 2000s.* New York: Russell Sage Foundation, 2013.

Kalleberg, Arne L. "Precarious Work, Insecure Workers: Employment Relations in Transition." *American Sociological Review* 74, no. 1 (2009): 1–22.

Kalleberg, Arne L. "Nonstandard Employment Relations: Part-Time, Temporary and Contract Work." *Annual Review of Sociology* 26, no. 1 (2000): 341–365.

Kalleberg, Arne L., and Michael Dunn. "Good Jobs, Bad Jobs in the Gig Economy." *Perspectives on Work* 20, no. 1 (2016): 10–13, 74–75.

Katz, James E., and Mark Aakhus, eds. *Perpetual Contact: Mobile Communication, Private Talk, Public Performance.* Cambridge, UK: Cambridge University Press, 2002.

Katz, James E., and Ronald E. Rice. *Social Consequences of Internet Use: Access, Involvement and Interaction.* Cambridge, MA: MIT Press, 2002.

Katz, Lawrence F., and Alan B. Krueger. *The Rise and Nature of Alternative Work Arrangements in the United States, 1995–2015.* Washington, DC: National Bureau of Economic Research, 2016.

Katz, Lawrence F., and Alan B. Krueger. "The Rise and Nature of Alternative Work Arrangements in the United States, 1995–2015." *ILR Review* 72 (2019): 382–416.

Katz, Vikki S., and Carmen Gonzalez. "Toward Meaningful Connectivity: Using Multilevel Communication Research to Reframe Digital Inequality." *Journal of Communication* 66, no. 2 (2016): 236–249.

Khan, Shamus Rahman. *Privilege: The Making of an Adolescent Elite at St. Paul's School.* Princeton, NJ: Princeton University Press, 2011.

Klaehne, Maurice. "Amazon Leads the Online Smartphone Sales Channel in the US in Q1 2018." *Counterpoint Research*, June 6, 2018. https://www.counterpointresearch.com/ amazon-leads-online-smartphone-sales-channel-us-q1-2018/.

Kopf, Dan. "Slowly but Surely, Working at Home Is Becoming More Common." *Quartz*, September 18, 2018. https://qz.com/work/1392302/more-than-5-of-americans-now-work-from-home-new-statistics-show/.

Kosanovich, Karen. "A Look at Contingent Workers." *US Bureau of Labor Statistics* 2018.

Kovacs, Anna-Maria. *Competition in the U.S. Wireless Services Market*. Washington, DC: Georgetown Center for Business and Public Policy, 2018.

Kumar, Krishan. *From Post-Industrial to Post-Modern Society: New Theories of the Contemporary World*. Oxford: Blackwell, 2005.

Kunda, Gideon. *Engineering Culture: Control and Commitment in a High-Tech Corporation*. Philadelphia: Temple University Press, 2006.

Labaton, Stephen. "New FCC Chief Would Curb Agency Reach." *New York Times*, February 7, 2001. https://www.nytimes.com/2001/02/07/business/new-fcc-chief-would-curb-agency-reach.html.

Lamont, Michele. *The Dignity of Working Men*. Berkeley: University of California Press, 2000.

Lane, Carrie M. *A Company of One: Insecurity, Independence, and the New World of White-Collar Unemployment*. Ithaca, NY: ILR Press, 2011.

Lane, Jeffrey. *The Digital Street*. New York: Oxford University Press, 2018.

Law, Pui-lam, and Yinni Peng. "Mobile Networks: Migrant Workers in Southern China." In *Handbook of Mobile Communications Studies*, edited by James. E. Katz, 55–65. Cambridge, MA: MIT Press, 2008.

Lenhart, Amanda. "Is Gig Work Our New Normal? Maybe Not." *Slate Magazine*, June 12, 2018.

Levy, Karen, and Solon Barocas. "Refractive Surveillance: Monitoring Customers to Manage Workers." *International Journal of Communication* 12 (2018): 1166–1188.

Lewis, Jamie. *Handheld Device Ownership: Reducing the Digital Divide? SEHSD Working Paper* 04. Washington, DC: US Census Bureau, 2017.

Licoppe, Christian. "'Connected' Presence: The Emergence of a New Repertoire for Managing Social Relationships in a Changing Communication Technoscape." *Environment and Planning D: Society and Space* 22, no. 1 (2004): 135–156.

Lifsher, Marc. "More Cellphone Users Switch to Prepaid Plans." *Phys.org*, February 22, 2013. http://phys.org/news/2013-02-cellphone-users-prepaid.html.

Ling, Richard. *Taken for Grantedness: The Embedding of Mobile Communication into Society*. Cambridge, MA: MIT Press, 2012.

Liu, Alan. *The Laws of Cool*. Chicago: University of Chicago Press, 2004.

Madden, Mary. *Privacy, Security, and Digital Inequality*. New York: Data & Society Research Institute, 2017.

Madden, Mary, Michele Gilman, Karen Levy, and Alice Marwick. "Privacy, Poverty, and Big Data: A Matrix of Vulnerabilities for Poor Americans." *Washington University Law Review* 95, no. 1 (2017): 53–125.

Madianou, Mirca, and Daniel Miller. *Migration and New Media: Transnational Families and Polymedia*. New York: Routledge, 2012.

Martin, Martin. *Computer and Internet Use in the United States: 2018*. Washington, DC: US Census Bureau, 2021.

Marwick, Alice. *Status Update: Celebrity, Publicity, and Branding in the Social Media Age*. New Haven, CT: Yale University Press, 2013.

Marwick, Alice, and danah boyd. "I Tweet Honestly, I Tweet Passionately: Twitter Users, Context Collapse, and the Imagined Audience." *New Media and Society* 13, no. 1 (2010): 114–133.

Matsakis, Louise. "Carpenter v. United States Decision Strengthens Digital Privacy." *Wired*, June 22, 2018. https://www.wired.com/story/carpenter-v-united-states-supr eme-court-digital-privacy/.

Mazmanian, Melissa, Wanda J. Orlikowski, and JoAnne Yates. "The Autonomy Paradox: The Implications of Mobile Email Devices for Knowledge Professionals." *Organization Science* 24, no. 5 (2013): 1337–1357.

McIntosh, Peggy. "White Privilege and Male Privilege: A Personal Account of Coming to See Correspondences Through Work in Women's Studies." 1988. https:// nationalseedproject.org/Key-SEED-Texts/white-privilege-and-male-privilege

McPherson, Tara. "Why Are the Digital Humanities So White? Or Thinking the Histories of Race and Computation." In *Debates in the Digital Humanities*, edited by Matthew Gold. Minneapolis: University of Minnesota Press, 2012.

Mears, Ashley. "Working for Free in the VIP: Relational Work and the Production of Consent." *American Sociological Review* 80 (2015): 1099–1122.

Merton, Robert K. *Social Theory and Social Structure*. New York: Free Press, 1957.

Mesch, Gustavo. "Minority Status and the Use of Computer-Mediated Communication: A Test of the Social Diversification Hypothesis." *Communication Research* 39, no. 3 (2012): 317–337.

Miller, Michelle, and Sam Adler-Bell. *The Datafication of Employment*. New York: The Century Foundation, 2018.

"Mobile Share of Advertising Market to Exceed 30% in 2020." *Zenith*, July 20, 2018. https://www.zenithmedia.com/insights/global-intelligence-issue-06-2018/mobile-share-of-advertising-market-to-exceed-30-in-2020/.

Moen, Phyllis, Jack Lam, Samantha Ammons, and Erin Kelly. "Time Work by Overworked Professionals: Strategies in Response to the Stress of Higher Status." *Work and Occupations* 40, no. 2 (2013): 79–114.

Morris, Seren. 2021. "Super Bowl Ad Cost 2021." *Newsweek*.

Mossberger, Karen, Caroline J. Tolbert, and Christopher Anderson. "The Mobile Internet and Digital Citizenship in African-American and Latino Communities." *Information, Communication & Society* 20, no. 10 (2017): 1587–1606.

Mossberger, Karen, Caroline J. Tolbert, and Mary Stansbury. *Virtual Inequality: Beyond the Digital Divide*. Washington, DC: Georgetown University Press, 2003.

Napoli, Philip, and Jonathan Obar. "The Emerging Mobile Internet Underclass: A Critique of Mobile Internet Access." *The Information Society* 30, no. 5 (2014): 323–334.

Neff, Gina. *Venture Labor: Work and the Burden of Risk in Innovative Industries*. Cambridge, MA: MIT Press, 2012.

Negroponte, Nicholas. *Being Digital*. New York: Alfred A. Knopf, 1995.

Nippert-Eng, Christena. *Islands of Privacy: Selective Concealment and Disclosure in Everyday Life*. Chicago: University of Chicago Press, 2010.

Omi, Michael, and Howard Winant. *Racial Formation in the United States: From the 1960s to the 1990s*. 2nd ed. New York: Routledge, 1994.

O'Reilly, Charles A., and Jennifer A. Chatman. "Culture as Social Control: Corporations, Cults, and Commitment." In *Research in Organizational Behavior: An Annual Series of Analytical Essays and Critical Reviews, Vol. 18*, edited by Barry Staw, 157–200. Boston: Elsevier Science/JAI Press, 1996.

Orlikowski, Wanda J., and Susan V. "The Algorithm and the Crowd: Considering the Materiality of Service Innovation." *MIS Quarterly* 39 (2015): 201–216.

Osterman, Paul. *Securing Prosperity: The American Labor Market: How It Has Changed and What to Do about It*. Princeton, NJ: Princeton University Press, 2000.

Parks, Lisa. "Around the Antenna Tree: The Politics of Infrastructural Visibility." *Flow*, March 5, 2010. https://www.flowjournal.org/2010/03/flow-favorites-around-the-ante nna-tree-the-politics-of-infrastructural-visibilitylisa-parks-uc-santa-barbara/.

Pearce, Katy E., and Ronald E. Rice. "Digital Divides from Access to Activities: Comparing Mobile and Personal Computer Internet Users." *Journal of Communication* 63, no. 4 (2013): 721–744.

Perlow, Leslie. *Sleeping with Your Smartphone: How to Break the 24/7 Habit and Change the Way You Work*. Cambridge, MA: Harvard Business Review Press, 2012.

Perlow, Leslie. "Boundary Control: The Social Ordering of Work and Family Time in a High-tech Corporation." *Administrative Science Quarterly* 43 (1998): 328–357.

Petriglieri, Gianpiero, Susan Ashford, and Amy Wrzesniewski. "Agony and Ecstasy in the Gig Economy: Cultivating Holding Environments for Precarious and Personalized Work Identities." *Administrative Science Quarterly* 64, no. 1 (2018): 1–47.

Pew Research Center. "Demographics of Internet and Home Broadband Usage in the United States." *Pew Research Center*, April 7, 2021. https://www.pewresearch.org/inter net/fact-sheet/internet-broadband/.

Pew Research Center. *Mobile Fact Sheet 2019*. Washington, DC: Pew Research Center, 2019.

Pew Research Center. *Mobile Technology and Home Broadband 2019*. Washington, DC: Pew Research Center, 2019.

Pew Research Center. "Demographics of Mobile Device Ownership and Adoption in the United States." Washington, DC, 2017. https://www.pewresearch.org/internet/fact-sheet/internet-broadband/.

Pinch, Trevor, and Wiebe Bijker. "The Social Construction of Facts and Artefacts? Or How the Sociology of Science and the Sociology of Technology Might Benefit Each Other." *Social Studies of Science* 14, no. 3 (1984): 399–441.

Pollert, Anna. *Girls, Wives, Factory Lives*. London: Macmillan, 1981.

Popiel, Pawel. "'Boundaryless' in the Creative Economy: Assessing Freelancing on Upwork." *Critical Studies in Media Communication* 34 (2017): 220–233.

Portwood-Stacer, Laura. "Media Refusal and Conspicuous Non-Consumption: The Performative and Political Dimensions of Facebook Abstention." *New Media & Society* 15, no. 7 (2012): 1041–1057.

Pugh, Allison. *The Tumbleweed Society: Working and Caring in an Age of Insecurity*. Oxford: Oxford University Press, 2015.

Pugh, Allison. "What Good Are Interviews for Thinking About Culture?" *American Journal of Cultural Sociology* 1 (2013): 42–68.

Pugh, Allison. *Longing and Belonging: Parents, Children, and Consumer Culture*. Berkeley: University of California Press, 2009.

Puhak, Janine. "Virginia Dunkin' Donuts Owner Calls Police on Black Customer for Using Free Wi-Fi without Purchase." *Fox News*, November 16, 2018. https://www.foxn ews.com/food-drink/dunkin-donuts-owner-calls-police-black-customer-using-wifi.

Putnam, Robert. 1995. *Bowling Alone: The Collapse and Revival of American Community*. New York: Touchstone Books, Simon and Schuster, 1995.

Qiu, Jack Linchuan. "Social Media on the Picket Line." *Media, Culture & Society* 38, no. 4 (2016): 619–633.

Qiu, Jack Linchuan. *Working-Class Network Society: Communication Technology and the Information Have-Less in Urban China*. Cambridge, MA: MIT University Press, 2009.

Qiu, Jack Linchuan. "The Wireless Leash: Mobile Messaging Service as a Means of Control." *International Journal of Communication* 1 (2007): 74–91.

Rafalow, Matthew H. *Digital Divisions: How Schools Create Inequality in the Tech Era.* Chicago: University of Chicago Press, 2020.

Rafalow, Matthew H. "Disciplining Play: Digital Youth Culture as Capital at School." *American Journal of Sociology* 123, no. 5 (2018): 1416–1452.

Ravenelle, Alexandrea. *Hustle and Gig.* Oakland: University of California Press, 2019.

Ritzer, George. *The McDonaldization of Society.* Thousand Oaks, CA: Pine Forge Press, 1993.

Robinson, Laura. "A Taste for the Necessary: A Bourdieuian Approach to Digital Inequality." *Information, Communication & Society* 12, no. 4 (2009): 488–507.

Robinson, Laura, Shelia R. Cotten, Hiroshi Ono, Anabel Quan-Haase, Gustavo Mesch, Wenhong Chen, Jeremy Schulz, Timothy M. Hale, and Michael J. Stern. "Digital Inequalities and Why They Matter." *Information, Communication & Society* 18, no. 5 (2015): 569–582.

Rodino-Colocino, Michelle. "Laboring under the Digital Divide." *New Media & Society* 8, no. 3 (2006): 487–511.

Rosenblat, Alex. *Uberland: How Algorithms Are Rewriting the Rules of Work.* Oakland: University of California Press, 2018.

Rosenblat, Alex, and Luke Stark. "Algorithmic Labor and Information Asymmetries: A Case Study of Uber's Drivers." *International Journal of Communication* 10 (2016): 3758–3784.

Rosenthal, Carolyn J. "Kinkeeping in the Familial Division of Labor." *Journal of Marriage and Family* 47, no. 4 (1985): 965–974.

Roy, Donald. "Quota Restriction and Goldbricking in a Machine Shop." *American Journal of Sociology* 57, no. 5 (1952): 427–442.

Scheerder, Anique, Alexander van Deursen, and Jan van Dijk. "Determinants of Internet Skills, Uses and Outcomes. A Systematic Review of the Second- and Third-Level Digital Divide." *Telematics and Informatics* 34, no. 8 (2017): 1607–1624. doi: 10.1016/j.tele.2017.07.007.

Schieman, Scott, Melissa Milkie, and Paul Glavin. "When Work Interferes with Life: Work-Nonwork Interference and the Influence of Work-Related Demands and Resources." *American Sociological Review* 74 (2009): 966–988.

Scholz, Trebor. *Uberworked and Underpaid: How Workers Are Disrupting the Digital Economy.* Cambridge, UK: Polity, 2016.

Schor, Juliet. *After the Gig: How the Sharing Economy Got Hijacked and How to Win It Back.* Oakland: University of California Press, 2020.

Schradie, Jen. *The Revolution That Wasn't: How Digital Activism Favors Conservatives.* Cambridge, MA: Harvard University Press, 2019.

Schroeder, Ralph. "Being There Together and the Future of Connected Presence." *Presence: Teleoperators and Virtual Environments* 15, no. 4 (2006): 438–454.

Schwalbe, Michael, Sandra Godwin, Daphne Holden et al. "Generic Processes in the Reproduction of Inequality: An Interactionist Analysis." *Social Forces* 79 (2000): 419–452.

Scott, Eugene Scott. "Chaffetz Walks Back Remarks on Low-Income Americans Choosing Health Care over iPhones." *CNN.com.* http://www.cnn.com/2017/03/07/politics/jason-chaffetz-health-care-iphones/index.html. March 7, 2017.

Scott, James C. *Weapons of the Weak: Everyday Forms of Peasant Resistance.* New Haven, CT: Yale University Press, 1987.

Seamster, Louise. "Black Debt, White Debt." *Contexts* 18, no. 1 (2019): 30–35.

Seamster, Louise, and Raphaël Charron-Chénier. "Predatory Inclusion and Education Debt: Rethinking the Racial Wealth Gap." *Social Currents* 4, no. 3 (2017): 199–207. doi: 10.1177/2329496516686620.

Selwyn, Neil. "Reconsidering Political and Popular Understandings of the Digital Divide." *New Media & Society* 6, no. 3 (2004): 341–362.

Senft, Theresa M. "Microcelebrity and the Branded Self." In *A Companion to New Media Dynamics*, 346–354. Hoboken, NJ: Wiley Online Library, 2013.

Sennett, Richard. *The Craftsman*. New Haven, CT: Yale University Press, 2008.

Sennett, Richard, ed. *The Culture of the New Capitalism*. New Haven, CT: Yale University Press, 2007.

Sennett, Richard. *The Corrosion of Character: The Personal Consequences of Work in the New Capitalism*. New York: W. W. Norton & Company, 2000.

Sennett, Richard, and Jonathan Cobb. *The Hidden Injuries of Class*. Reprint ed. New York: W. W. Norton & Company, 1993.

Shade, Colette. What "The Hustle" Looks Like on Etsy in 2015. *Jezebel* 2015.

Shahani, Aarti. "What You Need to Know About Subprime Lending for Smartphones." *NPR.org*, December 22, 2014. https://www.npr.org/sections/alltechconsidered/2014/12/22/372467354/what-you-need-to-know-about-subprime-lending-for-smartphones.

Sharma, Sarah. *In the Meantime: Temporality and Cultural Politics*. Durham, NC: Duke University Press Books, 2014.

Short, James, and Fred Strodtbeck. *Group Processes and Gang Delinquency*. Chicago: University of Chicago Press, 1965.

Silva, Jennifer. *Coming Up Short: Working-Class Adulthood in an Age of Uncertainty*. New York: Oxford University Press, 2013.

Slaughter, Anne-Marie. "The Gig Economy Can Actually Be Great for Women." *WIRED*, October 23, 2015. https://www.wired.com/2015/10/unfinished-business-women-men-work-family/.

Smith, Aaron. *Gig Work, Online Selling and Home Sharing*. Washington, DC: Pew Research Center, 2016.

Smith, Aaron. *Searching for Work in the Digital Era*. Washington, DC: Pew Research Center, 2015.

Smith, Aaron. *US Smartphone Use in 2015*. Washington, DC: Pew Internet and American Life Project, 2015.

Smith, Vicki. *Crossing the Great Divide: Worker Risk and Opportunity in the New Economy*. Ithaca, NY: ILR Press, 2002.

Snyder, Benjamin H. *The Disrupted Workplace: Time and the Moral Order of Flexible Capitalism*. Oxford: Oxford University Press, 2016.

Spence, Lester K. *Knocking the Hustle: Against the Neoliberal Turn in Black Politics*. Santa Barbara, CA: Punctum Books, 2015.

Spradley, James P. *The Ethnographic Interview*. New York: Harcourt, 1979.

Standing, Guy. *The Precariat: The New Dangerous Class*. New York: Bloomsbury, 2011.

Standing, Guy. "Tertiary Time: The Precariat's Dilemma." *Public Culture* 25 (2013): 5–23.

Stanley, Jay. *ACLU White Paper—Temperature Screening and Civil Liberties During an Epidemic*. New York: American Civil Liberties Union, 2020.

Star, Susan Leigh. "Ethnography of Infrastructure." *American Behavioral Scientist* 43, no. 3 (1999): 377–391.

Star, Susan Leigh, and A. Strauss. "Layers of Silence, Arenas of Voice: The Ecology of Visible and Invisible Work." *Computer Supported Cooperative Work: The Journal of Collaborative Computing* 8 (1999): 9–30.

Stempel, Jonathan. "New York City Sues T-Mobile over 'Rampant' Customer Sales Abuses." *Reuters*, September 5, 2019. https://www.reuters.com/article/us-t-mobile-us-new-york-lawsuit-idUSKCN1VP30N.

Stephens, Keri K. 2018. *Negotiating Control: Organizations and Mobile Communication.* New York: Oxford University Press.

Stewart, Gary. 1997. "Black Codes and Broken Windows: The Legacy of Racial Hegemony in Anti-Gang Civil Injunctions." *Yale Law Journal* 107, no. 7 (2017): 2249–2280.

Stuart, Forrest. *Ballad of the Bullet: Gangs, Drill Music, and the Power of Online Infamy.* Princeton, NJ: Princeton University Press, 2020.

Sugie, Naomi. "Work as Foraging: A Smartphone Study of Job Search and Employment after Prison." *American Journal of Sociology* 123, no. 5 (2018): 1453–1491.

Sundararajan, Arun. *The Sharing Economy: The End of Employment and the Rise of Crowd-Based Capitalism.* Cambridge, MA: MIT Press, 2016.

Swartz, David. *Culture and Power: The Sociology of Pierre Bourdieu.* Chicago: University of Chicago Press, 1998.

Swidler, Anne. "Culture in Action." *American Sociological Review* 51 (1986): 273–286.

Taylor, Keeanga-Yamahtta. *Race for Profit: How Banks and the Real Estate Industry Undermined Black Homeownership.* Chapel Hill: University of North Carolina Press, 2019.

Terranova, Tiziana. "Free Labor: Producing Culture for the Digital Economy." *Social Text* 63 (2000): 33–58.

Thrift, Nigel J. *Knowing Capitalism.* London: Sage, 2005.

Thompson, Paul, and Chris Smith. *Working Life: Renewing Labour Process Analysis.* New York: Palgrave Macmillan, 2010.

Ticona, Julia. "Strategies of Control: Workers' Use of ICTs to Shape Knowledge and Service Work." *Information, Communication & Society* 18, no. 5 (2015): 509–523.

Ticona, Julia, and Andrew Selbst. "In Carpenter Case, Supreme Court Must Understand that Cell Phones Aren't Voluntary." *Wired*, November 29, 2017. https://www.wired.com/story/supreme-court-must-understand-cell-phones-arent-optional/#:~:text=Today%20the%20Supreme%20Court%20will,location%20data%20without%20a%20warrant.&text=Therefore%2C%20the%20only%20action%20that,or%20using%20a%20cell%20phone.

Tomer, Adie, and Lara Fishbane. "How Cleveland Is Bridging Both Digital and Racial Divides." *Brookings* 2020.

Tomer, Adie, Lara Fishbane, Angela Siefer, and Bill Callahan. *Digital Prosperity: How Broadband Can Deliver Health and Equity to All Communities.* Washington, DC: Brookings Institute, 2020.

Towers, Ian, Linda Duxbury, Christopher Higgins, and John Thomas. "Time Thieves and Space Invaders: Technology, Work and the Organization." *Journal of Organizational Change Management* 19, no. 5 (2006): 593–618.

Troianovski, Anton. "The Web-Deprived Study at McDonald's." *Wall Street Journal*, January 29, 2013. http://www.wsj.com/articles/SB10001424127887324731304578189794161056954.

Tsetsi, Eric, and Stephen A. Rains. "Smartphone Internet Access and Use: Extending the Digital Divide and Usage Gap." *Mobile Media & Communication* 5, no. 3 (2017): 239–255.

Turkle, Sherry. *Alone Together: Why We Expect More from Technology and Less from Each Other.* New York: Basic Books, 2011.

Turner, Fred. *From Counterculture to Cyberculture: Stewart Brand, the Whole Earth Network, and the Rise of Digital Utopianism.* Chicago: University of Chicago Press, 2006.

Vallas, Steven P., and Angele Christin. "Work and Identity in an Era of Precarious Employment: How Workers Respond to 'Personal Branding' Discourse." *Work and Occupations* 48 (2018): 3–37.

van Deursen, Alexander J. A. M., and Jan A. G. M. van Dijk. "The Digital Divide Shifts to Differences in Usage." *New Media & Society* 16, no. 3 (2014): 507–526. doi: 10.1177/1461444813487959.

van Dijk, Jan A. G. M. "Digital Divide Research, Achievements and Shortcomings." *Poetics* 34, no. 4 (2006): 221–235.

van Dijk, Jan A. G. M. *The Deepening Divide: Inequality in the Information Society.* Thousand Oaks, CA: Sage, 2005.

van Doorn, Niels. "Platform Labor: On the Gendered and Racialized Exploitation of Low-Income Service Work in the 'on-Demand' Economy." *Information, Communication & Society* 20, no. 6 (2017): 1–17.

Van Oort, Madison. "The Emotional Labor of Surveillance: Digital Control in Fast Fashion Retail." *Critical Sociology* 45, no. 7–8 (2018): 1167–1179.

Venkatesh, Sudhir. *Off the Books: The Underground Economy of the Urban Poor.* Cambridge, MA: Harvard University Press, 2009.

Venkatesh, Sudhir. "'Doin' the Hustle' Constructing the Ethnographer in the American Ghetto." *Ethnography* 3 (2002): 91–111.

Wacquant, Loïc. "Inside the Zone: The Social Art of the Hustler in the Black American Ghetto." *Theory, Culture, and Society* 15 (1998): 1–36.

Wajcman, Judy. *Pressed for Time: The Acceleration of Life in Digital Capitalism.* Chicago: University of Chicago Press, 2014.

Wallis, Cara. *Technomobility in China: Young Migrant Women and Mobile Phones.* New York: New York University Press, 2013.

Warschauer, Mark. *Technology and Social Inclusion: Rethinking the Digital Divide.* Cambridge, MA: MIT Press, 2004.

Webster, Edward, Rob Lambert, and Andries Bezuidenhout. *Grounding Globalization: Labour in the Age of Insecurity.* Oxford: Blackwell, 2008.

Weil, David. *The Fissured Workplace: Why Work Became So Bad for So Many and What Can Be Done to Improve It.* Cambridge, MA: Harvard University Press, 2017.

Wellman, Barry, Anabel Quan Haase, James Witte, and Keith Hampton. "Capitalizing on the Internet: Social Contact, Civic Engagement, and Sense of Community." In *The Internet in Everyday Life,* edited by Caroline Haythornthwaite and Barry Wellman, 436–455. Malden, MA: John Wiley and Sons, 2002.

Wenger, Etienne. *Communities of Practice: Learning, Meaning, and Identity.* 1st ed. Cambridge, UK: Cambridge University Press, 1999.

West, Candace, and Sarah Fenstermaker. "Doing Difference." *Gender & Society* 9, no. 1 (1995): 8–37. doi: 10.1177/089124395009001002.

Wiatrowski, Bill. "BLS Measures Electronically Mediated Work." *Commissioner's Corner,* September 28, 2018. https://beta.bls.gov/labs/blogs/2018/09/28/bls-measures-electronically-mediated-work/.

Williams, Christine, and Catherine Connell. "Looking Good and Sounding Right." *Work and Occupations* 37 (2010): 349–377.

Wilson, William Julius. *When Work Disappears: The World of the New Urban Poor*. New York: Vintage, 1997.

Wimmer, Andreas. "The Making and Unmaking of Ethnic Boundaries: A Multilevel Process Theory." *American Journal of Sociology* 113, no. 4 (2008): 970–1022.

Wood, Alex J., Vili Lehdonvirta, and Mark Graham. "Workers of the Internet Unite? Online Freelancer Organisation among Remote Gig Economy Workers in Six Asian and African Countriew." *New Technology, Work and Employment* 33, no. 2 (2018): 95–112.

Wright, Erik Olin, Cynthia Costello, David Hachen, and Joey Sprague. "The American Class Structure." *American Sociological Review* 47, no. 6 (1982): 709–726. doi: 10.2307/2095208.

Yang, Tian, Julia Ticona, and Yphtak Lelkes. "Policing the Digital Divide: Institutional Gate-Keeping and Criminalizing Digital Inclusion." *Journal of Communication* 71 (2021): 572–597.

Zerubavel, Eviatar. "Private Time and Public Time: The Temporal Structure of Social Accessibility and Professional Commitments." *Social Forces* 58, no. 1 (1979): 38–58.

Zillien, Nicole, and Eszter Hargittai. "Digital Distinction: Status-Specific Types of Internet Usage." *Social Science Quarterly* 90, no. 2 (2009): 274–291. doi: 10.1111/j.1540-6237.2009.00617.x.

Zuboff, Shoshana. *In The Age of the Smart Machine: The Future of Work and Power*. Reprint ed. New York: Basic Books, 1989.

Index

For the benefit of digital users, indexed terms that span two pages (e.g., 52–53) may, on occasion, appear on only one of those pages.

Figures are indicated by *f* following the page number